DEDICATION

To Glynn Christian, the first to make grains interesting, and still a constant source of inspiration.

A cook's guide to grains

Jenni Muir

A cook's guide to grains

delicious recipes, culinary advice and nutritional facts

special photography by Jason Lowe

conran
OCTOPUS

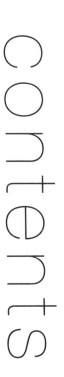

contents

Let's look at the menu. Quinoa, barley wheat berries, wild rice, brown rice and polenta. On the shelves there is organic soba from Japan, Forbidden Rice from China, and a range of unusual flours. This has got to be some hippy café run by an earnest, incense-burning, sandal-wearing, beard-growing, meat-avoiding Aquarian. No, actually, it's a traiteur, and it's owned by top American chef Charlie Trotter.

introduction

At Trotter's To Go in the Lincoln Park/DePaul area of Chicago, fresh takeaway food of outstanding provenance is prepared daily according to what is in season. The quinoa may come with roasted yams and spit-roasted turkey, the barley perhaps in a salad of roast aubergine and fennel, the wild rice stirred into a mixture of curried chicken and apricots. Then there is yellow corn grits with mushrooms and white truffle oil, or polenta flavoured with Maytag Blue cheese, to accompany slow-braised and spit-roasted meats. The spin: here's all the high-quality gourmet food to be found at world-renowned Charlie Trotter's eponymous restaurant, but it's supplied for your table, whether that's in the office, or at home.

Trotter's To Go reveals much about the way we want to eat today. There are many times after a hard day at work when we simply don't want to cook, yet the more we are exposed to fine restaurants, the more we expect all the food we eat to be of restaurant calibre. We are also increasingly concerned about health and the provenance of the food we eat, and the more frequently we put our meals in the hands of other suppliers, whether restaurant- or retail-based, the more they are required to provide not just food but also nourishment. In short, we want our day-to-day food, whoever cooks it, to be delicious and healthy.

It's no accident that grains and grain products feature so predominately at Trotter's To Go. Quite apart from the fact that Charlie Trotter enjoys using them (and fruit and vegetables) more than many of his un-contemporary contemporaries, the general public is steadily embracing the message that grain foods – preferably those that are not heavily processed – are an important base of a healthy diet.

Grains are upwardly mobile, and today interesting speciality varieties are more likely to be found in top restaurants, gourmet stores and via funky electronic retailers than in their typical previous home of the natural health store. In Western countries for the past 30 or 40 years, grains have been associated primarily with a puritanical style of eating that put perceived worthiness above pleasure. For centuries prior to that grains were associated with the peasant classes, and consequently thought to be food for the ignorant poor. They did pretty well on them, thanks, but the inherent nature of society at that time (and to a certain extent it remains so today) was for people in rural areas with little material wealth to mimic urban sophistication. As long as grains, and especially unrefined wholegrains, were perceived as a marker of poverty and ignorance, they would be cast aside as people gradually developed the means of affording rich people's food.

Now we have come full circle. The wealthy and famous eat grains because they want to stay healthy and live longer, and to show off the sophistication gleaned on international travels by embracing dishes from around the world, and for once the rest of us could do well to follow their lead.

Early man's domestication of wheat and barley from their wild, grassy ancestors formed the basis of civilization. Academics differ on whether this was the dawn of our glorious modern society, or the point at which it all went horribly wrong. Before we learnt to collect seeds and plant them we were hunter-gatherers, and in winter and early spring, food could be hard to come by. Collecting and storing food for consumption during the cold, lean months was therefore vital to survival, hence the attraction of growing grain. Although it was fiddly to collect and process, it could be stored far longer than fresh meat, fruits or leaves and also provided warming satiety when cooked. Grains have been the mainstay of the human diet for literally thousands of years. Though some argue that a more natural way of eating would be to consume only meats, nuts, roots and berries, there are few among us prepared to

swap our current pinnacle of civilization – a cosy evening watching *Frasier* on television – for chasing wild boar around with a spear.

In any case, the current debate about protein versus carbohydrates misses the mark. Many people thrive on a diet high in lean protein, just as others thrive on a diet rich in grains, fruit and vegetables. The real issue is that Western society needs to reduce its intake of highly processed refined white starches, sugar and hydrogenated fats, and is being given very little assistance in doing so. Disturbingly ignorant food and medical professionals exhort us to 'eat more wholegrains like couscous', when the couscous sold in stores is neither grain, nor whole.

So what are grains? That's a tough question. It's easier to point to several things that they are not, though very many things that are not grain – beans, pulses, legumes and some seeds – are also good nutritious foods. The trusty *Collins English Dictionary* defines grain as 'the small hard seedlike fruit of a grass, especially a cereal plant', as well as 'a mass of such fruits, especially when gathered for food' and 'the plants, collectively, from which such fruits are harvested'. Cereal, meanwhile, means 'any grass that produces an edible grain, such as oat, rye, wheat, rice, maize, sorghum and millet'.

For the purpose of this book, the definition is expanded to include some other seeds, such as buckwheat, quinoa and amaranth, that are treated as grains in the kitchen, as well as a few grains that are cereals, but not typically included in the academic definition. My aim is to help you feel confident using a wide variety of grains. Information on the provenance of each grain, the myriad ways in which it is used authentically, plus fresh ideas from leading professional cooks and advice on basic culinary techniques, are combined to help you feel more comfortable buying and using grains, especially unusual varieties, in your own kitchen and in your own ways. Please enjoy.

Grain farming

a profile

Crowmarsh Battle Farms near Benson in Oxfordshire is run by Philip Chamberlain, the third generation to run the family farming business.

Philip's mission is sustainable farming – to run a profitable business, generating a reasonable income, while maintaining and enhancing the farm environment and landscape. He advocates a system known in the industry as Integrated Farm Management, or IFM, a system that he has been knowingly practising for about five years, but on another level, following for much longer. Although not as well-known among the general public as it perhaps should be, IFM is thought to be the way forward by many who believe that it is not possible to feed the world organically, or that organic farming cannot be reconciled with society's demand for cheap food. IFM may be the ideal compromise.

It was his grandfather who decided it should be an arable farm 106 years ago. 'In this area we do not have the soil type for fruit and vegetables. It's too stony for sugar beets and potatoes, so root crops would not be economic.' The crops are combinable, meaning that they are harvested with a combine harvester, and they can all be processed using the same machine by adjusting its various settings.

The driver of the business is winter wheat – that is, wheat that is in the ground over winter – which is the most profitable grain. The farm works on a six-year rotation pattern. In the first year of the pattern, winter wheat will be planted. The following year, if the soil is good, there will be a second year of winter wheat or, if the soil is poor, a year of barley as it is easier to grow in poor ground. The third year will be an oil seed such as rape or linseed. Year four will be winter wheat, year five wheat or barley chosen as per year two, and then the six-year rotation will finish with a pulse or legume. 'Years one and four are potentially the best years for the wheat because then it is following a different crop,' says Philip.

For 800 hectares of winter wheat, eight varieties of the grain will be chosen. 'We don't want all of one variety in case it is prone to particular diseases or weather problems, and we do not plant more than 100 hectares of one variety. The factors affecting the output are the yield times and the price, and the yield and price both vary between varieties, and the price in particular varies according to the end use.'

Milling wheat for bread commands a premium price while wheat to be used for animal feed is the bottom of the market. The higher the value of the grain, the lower the potential yield. Crops for human consumption cost more to grow, partly because of the demand to look good and be free of stones.

Ease of marketing is also an issue. 'The end market depends very much on haulage limitations. In Scotland, for example, they can grow varieties of grain suitable for local distilling, but with the cost of haulage from Oxfordshire for us there is no point.'

Plant breeding improvements mean wheat varieties last only four to six years before being superseded, therefore the farm will often be growing one state-of-the-art variety that may become a key variety in a few years' time. 'XI-19 is a new milling wheat this year,' says Philip. 'It looks right and the trade may like it, but it also depends on whether the local mills want it.'

During spring they fertilize the autumn crops. 'That's where the art comes in, choosing the right chemical. First we identify the problem, then choose one according to what it will do and what it will cost.'

Sometimes chemicals won't even be necessary. If the pest is aphids, for example, there may be enough of their natural predators – ladybirds – to control them. But if there are not enough ladybirds, risks have to be assessed and decisions made. 'We have to consider the potential environmental impact – a cheap chemical may kill all the ladybirds as well as the aphids. Then we ask if we have to spray the whole field. It may be that only the microclimate near the Thames is affected and we need only spray a few metres around the outside. Then we play with the dosage rates.' If using one strength of chemical solution works, next time they will try to reduce it.

This decision-making process has to be used throughout the year to assess various problems. In general, they wait to see if a problem arises and then address it, unlike preventative prophylactic spraying, which works rather like a flu jab. If a farmer knows a field always has wild oat he may spray it every year no matter what. This may sound like a recipe for environmental disaster, however it can also be the case that the chemical used in prophylactic spraying is less damaging overall than chemicals that might be applied after a problem arises.

Satellite technology is used to analyze the fields so that chemicals are deposited only where necessary. Specific local knowledge also plays a key role. 'Part of a field may seem as though it needs more fertilizer but that may be because there is a bank of rabbits there that needs controlling.'

Looking after the environment on a farm involves things that can be seen: the hedgerows, trees, grass, birds and bees. However, IFM also encompasses less visible issues such as water pollution, and the global environment, in aiming to address the reliance upon non-renewable resources. 'By choosing different cultivation techniques we can affect the wear on non-renewable resources. For example, we can adjust the working depth in soils to minimize wear and tear on the machinery.'

The sewage that should no longer be dumped at sea or burned to pollute the atmosphere has found a positive role on the farm, which has a contract with the local water company to make use of it. 'It is an excellent fertilizer,' says Philip, 'and a very renewable resource.' Indeed, it's about as sustainable as we can possibly get.

Grain processing

Processing grains may well be one of the oldest professions in the world. As Tom Stobart says in his *Cook's Encyclopaedia*, 'since such creatures as different as otters and vultures have learned to use stones as tools, it is probable that man learned to pound grain even before he could grow it.'

Any inedible husk or hull surrounding the grain needs to be removed before the grain can be consumed (some varieties of grain have no husk at all). Then there are several layers of bran that may be stripped away from the grain. Below the bran is the aelurone layer, which covers the starchy endosperm. Secreted beneath this fat, protective layer is the seed of the plant known as the germ. A wholegrain is whole to the extent that it contains the bran, aelurone layer, endosperm and germ.

The grain may then be processed in various ways. Simple cracking, or chopping the grain into two or three pieces, exposes the grain's starchy endosperm and shortens cooking time. Examples include cracked wheat, cracked buckwheat and steel-cut or 'Scots' oats.

Alternatively the grain is pearled or polished. This process strips away the outer layers of bran to varying degrees. Grains that are semi-pearled, semi-perlato, or semi-integrale are effectively only partially stripped of bran and therefore retain more fibre and other nutrients. This shortens the cooking time to an extent; semi-pearled products need less time to cook than wholegrains, but more than fully pearled or polished grains. Far from being an uncomfortable compromise between 'worthy' wholegrains and 'highly processed' polished grains, semi-pearled products are often superior and arguably ideal. They are neither too heavy nor too light. They boast excellent nutty flavours and good nutritional benefits while at the same time moderating cooking times.

Continuing to strip away the bran layers of the grain results in polished grains, of which white rice is the most familiar. Thoroughly stripped, white pearled barley is popular and pearled wheat products such as Pasta Wheat and Ebly are also available.

Instead of pearling, the grain may be crushed and ground to varying degrees. Crushed grains are called grits or groats and are midway between cracked grains and fine flour. Semolina, which is the coarsely milled endosperm of the grain, is similar. Although it can be made from corn and rice, the term is especially associated with the durum variety of wheat. When cooked with liquid, semolina has a slightly rough rather than smooth paste-like texture.

Ground grains are often called meal and have an appealing texture. Apart from oatmeal, which comes in various grades, some of which are quite large, most products called meal tend to be like rough flour, with some tiny particles and some larger. Flours are created when the grain is ground to very small particles, but even these fine flours have varying degrees of texture, with some rated as superfine.

Wholemeal flours are made from the wholegrain, unlike most refined white flours that have had the bran and the germ removed. Removing these elements makes for a lighter flour but also prolongs their shelf life as the bran and germ contain oils that become rancid with time. Some flours have these elements removed during processing then added back in to create various degrees of 'browness'. These brown flours, however, are not in the spirit of the natural foods lifestyle as they are highly processed and not 'whole'.

The two key methods of milling flour are roller milling and stonegrinding. Roller milling is a fairly modern, somewhat controversial method. In addition to stripping away the bran and germ of the grain, it generates heat as it does so, which is thought to damage the enzymes in the flour. Many of the large factories that produce such flour also add chemical bleaching agents and 'improvers' to flour, which may include such tasty delights as benzoyl peroxide,

chlorine dioxide, ammonium and potassium persulsphates and more. These additives make the flour whiter and simulate the natural ageing process in quick time in order to produce breads of large volume and soft, fine texture.

Stonegrinding, in which, as the name suggests, the grains are rubbed between a set of stones, is the preferred method of flour production for many expert bakers and produces richer flours of great flavour and nutrition because the germ of the grain is retained. The flours also tend to be aged naturally so that good bread can be made from them without the addition of chemicals.

Roller mills are also capable of producing rolled flakes of grain that are most familiar as rolled oats, but the process is also used for wheat, barley and rye. The process involves pearling the grain, then steaming it so that it is soft enough to roll flat. Shelf life is prolonged because the heat destroys enzymes that would turn the oil in the grain rancid.

Some grains are roasted or parched during processing to produce various levels of flavour. Bere, for example, is lightly kiln-roasted before being ground, while wild rice is parched to create a nutty flavour and prolong storage life. Products such as freekeh, which is made from young green wheat, are roasted to a much greater extent to produce grains with an intense (and slightly smoky) flavour.

A wholegrain is malted when it has been soaked, allowed to sprout, and then dried and matured before milling. This process can produce anything from pale to very dark malt. It is used in brewing, making bedtime drinks, as a nutritional supplement and in cooking, especially porridges and baking. In general it adds a sweet, rich and slightly caramelized flavour.

The other key grain processing technique is alkali processing, which is restricted to corn. This creates hominy, sometimes known as hulled corn. Hominy is occasionally labelled pozole or posole.

Grains and Health

It is a truth universally acknowledged that a celebrity in possession of a large fortune must be in want of a nutrition guru. These happy marriages and the plethora of books, articles and magazines devoted to their promotion are often criticized for encouraging an impossible ideal of health among the general public, but historically, things have been a lot worse.

For centuries, society's poorer people have wanted to mimic the diets and lifestyles of the rich and aristocratic. The unfortunate result has been that, over time, the general public have become used to diets based on highly refined wheat flour, fat and sugar that at one time only the very wealthy could afford on a daily basis.

In the past 20 years or so we have come full circle. Now the wealthy think it elegant and stylish to eat rustic peasant food and pay handsomely for the privilege. Restaurants fly in wild herbs, organic meats, cold-pressed oils and speciality grains from unpolluted hillsides throughout the Mediterranean. Meanwhile the peasant's contemporary urban equivalent is malnourished on a diet high in fat, sugars and refined starches, and low in fresh, natural, minimally processed foods.

If the current obsession with celebrity has any long-term value to society at all, it may well lie in helping those who are not wealthy to connect poor health with negative eating and lifestyle habits. Our governments and medical services are increasingly encouraging us to do so, for it is now clear that many modern ailments and conditions are caused or exacerbated by poor eating and lifestyle habits, and the incidence of obesity is steadily increasing. If people reduce their intake of junk food because they see that some celebrity benefited from doing so, it is ultimately a good thing.

Grains are important sources of many nutrients, several of which are vital to good health. Key among these is fibre or 'roughage', which is only found in plant foods. It is recommended that we eat 30g/1oz of fibre per day. Fibre passes undigested through the stomach and small intestine. On a basic level, it 'keeps you regular', however it also has many less obvious health benefits. Soluble fibre, which dissolves in water to form a thick gel, keeps the bowel healthy, helps prevent cancer, lowers blood cholesterol levels and therefore helps reduce the risk of heart disease. It is especially helpful for diabetics because it slows the release of sugars in the blood. Insoluble fibre helps push soluble fibre through to the colon, prevents constipation and related problems such as haemorrhoids and diverticulitis. It is also thought to reduce the risk of colon cancer. Another type of fibre is resistant starch. Like insoluble fibre, it leaves the bowel undigested and is thought to reduce the risk of cancer, help control diabetes and encourage the growth of beneficial bacteria.

Wholegrain foods are higher in fibre than processed grains, which is why it is recommended that we eat some wholegrain foods daily. However, in recent years it has also been discovered that wholegrains contain other beneficial nutrients such as phytoestrogens that protect against some forms of cancer, heart disease and may relieve menopausal symptoms. The phytic acid found in the bran of the grain may protect against some cancers, while the phytosterols in cereal oils can reduce blood cholesterol.

To make dietary advice simpler to understand, grains have traditionally been categorized as carbohydrate foods, in particular complex carbohydrates. However, even if we set aside the high-protein, lysine-rich seed grains featured in this book and concentrate on the main cereal grasses – wheat, rye, barley, oats, rice, millet, sorghum and corn – it is apparent that grains are major sources of protein. Thinking of them only as carbohydrates oversimplifies the issue. If cereal grains were your sole source of food (which they should not be), you would be consuming more protein than the body requires and your fat intake would be undesirably low. On average, these eight cereal grain species provide 45 per cent of the world's dietary protein and 50 per cent of its total food energy (however depending on the specific region and the amount consumed, cereals provide 20 to 70 per cent of humans' dietary protein).

Cereal grains have been dismissed as valuable protein sources because of their generally low lysine content, making them what has popularly been termed 'incomplete' proteins. There is a growing appreciation, however, that this perceived 'lower biological value' of cereals compared to animal proteins is irrelevant. This is because the other foods eaten during the day – beans and pulses, nuts and seeds, meat and dairy products – compensate or complement the nutritional benefits of grains, making the total protein consumed 'complete'.

In any case, the more science learns about food, the more it becomes apparent that the old, narrow focus on the carbohydrate, protein and fat content of foods is limited in its helpfulness. Most important today is the recognition that for maximum nourishment we need to eat a wide variety of foods. Japan led the way in this respect years ago, advising the public that they should aim to eat 30 or more different types of food from across six food groups each day, while also balancing calorie intake with energy expenditure.

Only recently has similar emphasis on variety made it to the forefront of health advice given in the West, but it has had teething troubles. Some elements of the media have misinterpreted the variety message and promoted fear that eating a wider variety of foods will lead to weight gain. Eating a variety of foods does not mean consuming different bags of crispy corn snacks. It requires us to embrace the huge diversity of ingredients nature has to offer. And it means not eating the same ingredients two meals in a row, or even two days in a row.

There would perhaps be less confusion if people today were not so out of touch with cooking and the means by which food is produced. The message about variety will be better served when the food industry and general public learn to focus on ingredients, whether that be raw ingredients for cooking or the components of a ready-made meal.

Exploring the range of grains available to us – whole, ground into flours, or minimally processed into convenient foods such as pasta – is an easy way to begin eating a wider variety of foods. Too many people in the West eat a processed wheat breakfast cereal in the morning, a wheat bread sandwich for lunch, and spaghetti (more wheat) for dinner. Such a diet can be moderate in terms of fat and calories, but it does not incorporate a variety of foods or maximize nutrients. The simplest change for someone with such an eating pattern would be to switch to an oat-based breakfast cereal in the morning, and rice or rice noodles in the evening. But that is only a first step.

In addition to optimizing the range of nutrients we consume, eating a variety of foods minimizes the likelihood of developing or aggravating food intolerances, also known as food hypersensitivity. Dr Robert Atkins, a specialist in this field (but infamous for his low-carbohydrate weight-loss diet), emphasizes the importance of varying or 'rotating' the foods we consume. 'I think that the habit of eating certain foods every day is one you should take pains to avoid,' he says. 'In fact, I would recommend that you never have the same food on successive days.'

The Okinawa Centenarian Study, based on 25 years of research into the eating habits and lifestyle of the Okinawa islanders, the world's longest-living population, is arguably the most up-to-date, authoritative reference guide available for those interested in healthy eating. It suggests several adjustments to the food pyramid recommended by the United States Department of Agriculture (USDA) and similar organizations world wide, which many health professionals have come to see as inadequate. Dr Walter Willet, chair of the Department of Nutrition at Harvard School of Public Health, for example, has outlined several problems with the standard food pyramid, one of which is that it needs to put more emphasis on whole grains.

The Okinawa food pyramid has as its foundation the recommendation to eat ten servings of whole grains and ten servings of vegetables and fruit daily. In addition, it advises that we eat three servings of calcium foods (dairy, soy and some vegetables), three servings of 'flavonoid' foods (essentially soy products but also some fruit and veg), and two servings of Omega-3 foods (oily fish and linseeds). Further counsel is that tea and plenty of fresh water be drunk daily, while limiting meat, poultry and eggs to less than once a day, and consuming sweets no more than three times per week.

It is important to clarify that these serving sizes are relatively small and it is likely that the meals you currently eat contain more than one serving. A sandwich comprising two slices of bread would constitute two grain servings. If you followed it up with a generous scone you could be eating four grain servings. A large bowl of cereal – 50g/2oz dry weight, not the measly 25g/1oz they allow you at dieting clubs – would take care of another two grain servings, as would a whole bagel or two thin slices of toast. And there are very few adults who would think a half-cup of cooked rice (one serving) made a substantial portion at dinner, so it is easy to consume two or three servings in the evening too.

The people behind the study have not set the recommendations in stone either: there are recommended lower and upper limits for each category, and you would still be in line with their programme if you consumed as little as seven servings of grain foods daily. Apart from the challenge of adopting whole grain foods in place of refined products (even the Okinawans eat a large amount of polished white rice), most people today would find it more difficult to meet the targets for fruit and vegetable consumption than those for grains.

Many best-selling health books in recent years have focused on something called the Glycemic Index (GI). This is a system of measuring how different high-carbohydrate foods affect levels of blood sugar and insulin. It has a particular relevance to people with diabetes. In fact, there are two different GI systems – one based on glucose and one based on white bread. This alone should be an indication that attempting to follow either is too complicated for day-to-day healthy eating. In general, it is recommended that high GI foods are eaten in 'extreme moderation' because they provoke a strong insulin response in the body, and provide a quick burst of energy that quickly leads to further hunger. Low GI foods, on the other hand, provide a slow and sustained release of energy and satisfy hunger for longer periods.

The problem is ascertaining the GI of what you are eating in real life. For starters, the classification of various different foods is complicated. Both white and wholemeal breads have a high GI, while white pasta is rated as low GI. Even different varieties of polished rice have different GIs. An option might be to carry round a guide to the GI levels of various foods to refer to in restaurants and supermarkets, however you then also have to calculate the moderating effect on the GI of all foods consumed with your chosen carbohydrate food.

This is totally unworkable for the majority of the population and complicates the message we already know: that we would be better off eating unprocessed whole grain foods regularly. Maybe not always, but more often. Even the Ancient Greeks knew that high-fibre whole grain foods were beneficial for health and ate them as needed.

The Chinese and Indians both have ancient traditions of food-health that are experiencing greater credence in the West. They too consider grains to be essential health-promoting staples but specific recommendations are quite different from those given by Western medicine and nutrition practitioners. Chinese nutrition, for example, includes refined white rice and wheat products as the norm.

India's Ayurvedic system considers that whole, processed and refined grains all have a place in the diet and the choice at each meal should depend on the individual's body type, their current state of health, and the season. As Ayurvedic culinary expert Miriam Kasin Hospodar suggests in her book *Heaven's Banquet*, a key reason many health food advocates have only promoted unrefined whole grains in the past is in reaction to the widespread over-consumption of highly refined grains and grain products. 'The important things to look for are, first, as always, that the grains are organically farmed and, second, how and why they have been processed,' she says. 'Some grains are refined to make them more digestible, making their nutrients more accessible. Grains with a heavy husk, such as barley and rice, are hard to digest and their pearled counterparts are the better Ayurvedic choice.'

The lesson here is that there are many routes to health and it is essential to consider the needs of the individual and the degree of enjoyment that the food provides. Furthermore, good health is inextricably linked to lifestyle factors other than food. Of vital importance is regular activity and a feeling of social connection. Any eating programme that prevents you going out for a relaxing dinner with friends is not only likely to fail but is arguably unhealthy.

Good cooking begins with good shopping, but where does good shopping begin? With a sound general knowledge of ingredients: how to use them, why to use them, where to find them. Even our most basic staple foods have speciality varieties and uses. A simple bag of barley can take your tastebuds on a culinary journey from Morocco via Scotland to East Coast USA. Here's the travel brochure.

the grains

Wheat is
mild, sweet, chewy, crunchy, elastic, hearty, toasty

It goes with
chicken, nuts, mushrooms, dried fruit, cream, honey, dried ginger

Wheat
farro, spelt and kamut

Wheat has been having a tough time recently. Open any popular magazine or newspaper and some celebrity or another is attributing their new look/new vitality/new happiness to an avoidance of wheat and products made from it. Fast emerging as a pariah for the early twenty-first century, wheat is now about as welcome at elegant soirées as chicken pox.

The wheat- and flour-producing industries have mobilised their crisis management teams, funding studies and authorizing media information packs, quoting noted nutritionists and attempting to persuade both editors and the public that avoidance of carbohydrate-rich foods in general and wheat in particular is a bad idea.

In a press release distributed by Britain's Flour Advisory Bureau and headed 'Intolerance Myths Putting Women at Risk', Professor Tom Sanders, head of the Department of Nutrition and Dietetics at King's College London, is quoted as saying that unless someone suffers from the very rare condition of coeliac disease (a serious allergic reaction to gluten, a protein contained within wheat and some other cereal grains), cutting out wheat is extremely unwise.

'Many women believe they have a food allergy or intolerance, but in reality numerous studies have shown that only 1–2 per cent of the population suffer from a food intolerance and only 0.3 per cent suffer from Coeliac disease,' he says. 'The quality of a diet is all about what you include, not what you cut out.'

True, some people just want to give up certain foods and find a fashionable intolerance or allergy is a good excuse to do so. However, for the small group who suffer from wheat intolerance and associated conditions such as irritable bowel syndrome, life can be very uncomfortable. Avoiding wheat for a period, with a view to eventually consuming it again in limited quantities, can provide much relief. As this book demonstrates, providing you are happy to cook, it is perfectly possible to avoid wheat without cutting out all starchy foods, and one can do so while enjoying a delicious, highly nutritious and varied diet.

Alternatives to common wheat

An exciting facet of the fashion for wheat avoidance is the range of traditional or heirloom wheat varieties now available, and the research being done to establish whether these old-fashioned grains can be tolerated by people sensitive to common wheat.

Organic farmers Clare and Michael Marriage of Doves Farm Foods in Berkshire, England, have been instrumental in bringing spelt (*Triticum spelta*) to public attention and driving the market for interesting grain flours in Britain. 'Spelt is an ancient type of wheat that had fallen from favour but is now being grown in quite a few countries,' Clare explains. 'It has an inedible outer husk that common wheat doesn't have. Also, if you look at the DNA fingerprinting of different wheat varieties you can see slight differences. The banding of spelt is different from common wheat particularly in the gluten area. Spelt contains good-quality gluten, but the gluten fraction is very different.' This has the logistical effect of making spelt bread dough rise more vigorously than dough made from common wheat.

Petit épautre has been grown in the Haute Provence region of France for thousands of years. It is often described as a type of spelt; however it is not, and goes under the botanical name *Triticum monococcum*. (The French incidentally call spelt *le grand épautre*). Petit épautre has also been called einkorn. The French name is more romantic however, and it also suits the grain better: the elegantly shaped kernels are indeed petite, more like voluptuous pointed rice than most wholegrain wheat berries.

Today a syndicate of around 30 farmers is working to revive the popularity of this grain, which was a favourite in the Mediterranean region until Roman times. Farmed organically, petit épautre is often grown next to fields of lavender, a plant that also thrives in the Haute Provence's warm climate and chalky soil. It has an impressive nutritional profile, containing the eight essential amino acids, including lysine, which is often absent in other cereals. Just 100g/3½oz of petit épautre supplies the average person with a whole day's protein requirement. It also has a low level of gluten.

Farro (*Triticum dicoccum*) is another fashionable old wheat whose image has improved since dumping its unglamorous English name, emmer. Both emmer and einkorn are mentioned in classical Greek texts. Now a favourite of stylish Italian restaurants, farro is often made into soups with vegetables and pulses. Some food writers have described it as being lighter and more elegant than other wheats. This impression is not caused by the grain itself but the fact that it is usually sold semi-integrale, so has less bran on the kernel than wholegrain wheat berries do.

The taxonomy of the large-kernel wheat sold under the trademarked brand name Kamut has been disputed by scientists, however most now believe it is *Triticum turgidum turanicum*. Developed for the modern market by Montana-based Bob Quinn, an organic farmer, agricultural scientist and plant biochemist, it is related to durum wheat, the hard variety properly used for making pasta. The kernels are twice the size of common wheat and offer 20–40 per cent more protein as well as being high in gluten.

A study by the International Food Allergy Association has found that seven out of ten people who react negatively to common wheat are able to tolerate Kamut brand wheat when it is eaten as part of a rotation diet. This welcome research highlights the need for further scientific investigation of alternative wheats, especially in the light of the mounting anecdotal evidence that many wheat-intolerant people are able to consume spelt without experiencing negative reactions. (Typical reactions to common wheat might include abdominal cramps, fatigue, headaches, joint pain, breathing problems, sinus problems, hives and nausea or indigestion.)

More good news about these alternative wheats is that they can with relative ease be manufactured into all the wheat products people currently enjoy: breads, pasta, burghul, couscous, cakes and so on. Some commercial bakers cheekily label spelt products as being 'wheat free': although this is not the case, some wheat-intolerant people may well find they are able to consume them. The flours and ready-prepared foods made with those grains that are currently available tend to be of the wholegrain variety, primarily because it is the health food market that has shown most interest in alternative wheats. However there is no reason why refined products could not be made from them should a company care to produce the necessary flour.

Freekeh – the rich roast

When it comes to wheat, it seems everything old is new again. In South Australia, businessman Tony Lutfi and colleagues have developed a modern means of manufacturing freekeh (pronounced free-ka), a traditional and highly esteemed grain food of the Middle East, produced by roasting young spears of green wheat or barley to give a rich, smoky and somewhat meaty taste. In essence, the process is not very dissimilar to the manufacture of wild rice (see pages 46-49).

The legend is that, around 2300BC, the inhabitants of a walled city near the current border of Syria and Turkey expected an enemy to march against them. In a desperate effort to protect their crops, they harvested their wheat before it was ripe and brought it within the city walls, stacking it in a corner. The temperature of the young wheat increased while piled up, causing the heads to wilt and the husks to

dry. Eventually the stacks of wheat combusted and the people ran to save the grain from the fire. They started rubbing the damaged wheat and to their surprise found that the green grains did not burn, only the husks. And the result tasted wonderful. In the Armaic language (the parent language of Arabic and Hebrew) freekeh means 'the rubbed one'.

People have been producing freekeh in Syria, Jordan, Northern Iraq, Egypt, parts of Turkey and North Africa (where it is made with barley) ever since, using the basic method discovered by accident in the ancient walled city. The young green grain is harvested by hand, the heads tied together and laid on the ground in the midday heat so that it becomes dry enough to burn. Then they set fire to it.

Unfortunately with this method the freekeh also picks up a lot of dirt and stones. Putting the grains through a thresher only breaks up the stones so that they become even more difficult to distinguish from the grain, nor does winnowing help. The onus is then on the cook to sift through the freekeh with extreme care to remove all the stones so that none of the diners breaks a tooth. This is one reason why you may never have heard of freekeh, and why it has remained a rare speciality of the Middle East.

Tony Lutfi was introduced to freekeh while working in Jordan and dining at a wealthy friend's house. His host remarked that he was lucky to eat freekeh at his home because when served anywhere else it was likely to break his teeth. 'He said to me that if anyone could find a way to sell freekeh free from stones they would make a lot of money.'

Ever the businessman, Lutfi noted the comment with interest, but forgot about freekeh until dining at an Armenian house in Sydney where he was served it again and he then resolved to develop a means of producing it cleanly. One challenge was producing a new machine that could harvest green grain – by necessity this means that Lutfi takes his machine to

the farms rather than relying on farmers to bring their grain to him. Another key was devising a method whereby the harvested wheat does not come into contact with the ground at all.

'We haven't invented freekeh, but we have invented a method of producing it cleanly and developed what is the only technology of its kind in the world,' says Lutfi. 'Freekeh was the last food in the world that was so basic and yet undiscovered. We have turned it into a food for modern living with modern processing.' Production has doubled every year for the past four years and he exports to Jordan as well as other countries. In time he plans to produce freekeh made from barley and triticale as well as the durum wheat currently used.

Nutritionally, freekeh is very high in usable protein and fibre, so that while a normal serving of white rice might be 90g/3¼oz, a 50g/2oz serving of freekeh will provide the same degree of satiety. Therefore, while it may seem more expensive, the cost-per-serve is on par with white rice. 'Mature wheat has a high level of starch, 5–6 per cent fibre and a little sugar, however freekeh has a little starch, a little sugar and 16–17 per cent fibre,' says Lutfi. 'That's why freekeh tastes nicer – it is not starchy, but has a crunchy, nutty taste because of its high fibre content. Even brown rice only has 3.9 per cent fibre,' he adds, 'and freekeh is much higher in calcium and iron than any rice or pasta.'

There is also some anecdotal evidence to suggest that people intolerant to wheat can consume freekeh and Lutfi, aware of the potential marketing benefits, is exploring the possible reasons for this.

Adelaide has taken freekeh to its heart. Sold in whole and cracked form, it is the basis of the best-selling salad at popular café and bakery The Queen of Tarts and is elsewhere being made into breads. Lutfi advises 30 per cent freekeh flour to 70 per cent wheat flour is the ideal ratio for baking, however the whole or cracked grains can simply be kneaded into doughs. Australian chefs and food writers have got behind it too, notably Christine Mansfield who recommends freekeh for silky textured noodles. It is well received when made into burgers, pilaffs and stuffings. Lutfi is also looking at ways to employ this highly nutritious grain in breakfast cereals.

Wheatgrass – a green elixir

After a busy 12-hour day, Australian Rhonda Sweetgrass goes to work in her back garden performing the hard graft element of her business, planting wheat seeds. Rhonda will work for another six hours today, and the labour is physically demanding. She constantly bends and stretches, lifting and carrying heavy trays, spreading out the grains with her hands and stamping them down with her feet.

It is impossible not to muse that here is a slim woman in her fifties, confidently wearing a sleeveless top and tiny shorts, exuding a level of vitality and enthusiasm that would make most teenagers feel old and weary. Rhonda is extraordinarily focused and committed. To watch her is to think: 'Whatever she's on, I want some.'

What she's on, and what she's growing, is wheatgrass. The young blades of grass, propagated from wheat grains, are harvested when 15–25cm/6–10in tall and just developing a second stem, usually after nine to twelve days. At this point they are at a nutritional peak, boasting an impressive level of all minerals, especially calcium and zinc, and all B-vitamins, including folic acid and B12, plus a broad spectrum of other vitamins and phytonutrients. Wheatgrass is also, at this youthful stage, a complete source of protein.

When put through a specially designed juicer, a thick, intense green liquid is extracted that works on the body like a fresh natural vitamin supplement. This is drunk, ideally, first thing in the morning, just 30 ml/2 tbsp, taken like a nutrient-packed shot of espresso, and followed by a large glass of water. The flavour takes a little getting used to, but this is made easier if you are one of Sydney-based Rhonda's lucky customers. She isn't called Sweetgrass for nothing. The wheatgrass she works so hard to produce to meticulous organic and biodynamic standards is the best-tasting you can buy and is the result of much trial and experimentation, with the specific grains, with the growing method, and even with the water.

'I know what it's done for me,' she says, 'and I've never met someone who has taken wheatgrass and said they haven't benefited. Some people report feeling better immediately, for others it takes two or three days. Often people will take it for two weeks then decide they're over it and stop. I get a lot of calls from people who have stopped and noticed the difference – that's when they buy a wheatgrass juicer and start doing it themselves."

Rhonda is not a purist – she drinks around four cups of coffee a day, eats whatever she likes and has been known to smoke and drink alcohol – so on the occasions she is adamant about the finer points of taking wheatgrass, one pays attention. She doesn't believe it should be mixed, as many juice bars do, with vegetable and fruit juices – especially not with fruit. 'Purists hate the fact that I bottle it,' she admits. 'It is best when it is first juiced but for many people, if they couldn't get it in bottles, they wouldn't get it at all as many shops won't invest in the juicer and stock.'

The cost of the juicer is an issue for anyone wanting to DIY wheatgrass. Rhonda and several other experts favour the powerful, but high-priced Green Life juicer that has a twin-gear system that slowly grinds the maximum amount of juice from the grass while keeping nutritional qualities at their peak.

There are slightly cheaper juicers available, including hand-cranked machines, but these are not very good value. The regular centrifugal domestic juicers can't handle wheatgrass, despite what the manufacturers may claim, and certainly not in any quantity. Indeed, the more you look into it, the more you are likely to find you have no choice but to buy the best available.

Steve Meyerowitz, sprouting expert and author of *Wheatgrass: Nature's Finest Medicine* (required reading for anyone interested in the subject), has calculated the costs of growing and juicing wheatgrass yourself and compared it to purchasing the juice from stores. Once the inconvenience factor is taken into account, it works out about the same.

Wheat is not the only grain capable of producing grass for juicing. Some people believe barleygrass is even better nutritionally. 'It's very good for the system and has slightly more protein,' says Rhonda. 'It's also a bit easier to grow because it's not as temperamental. However it tastes bitter. We do a blend of barley and wheatgrasses in summer when the wheat is sweet enough to combat the taste of the barleygrass.' Similar juices can also be made from oat and rye grasses.

Rhonda is becoming used to the diversity of people enthusiastic about wheatgrass. In addition to organic retailers and juice bars, her clients include one of Sydney's top fine food stores as well as a kebab shop. 'There's a takeaway in Bondi called Hot and Juicy,' says Rhonda. 'It's the last place you'd expect to find wheatgrass but they sell it and call it Green Viagra. They know from their own experience that it works really well.'

But it is Rhonda and her attitude that is perhaps most persuasive. She is happy for sceptics to remain sceptical, and advises that the best way to decide for yourself about wheatgrass is to maintain any unwholesome habits. 'Stay unhealthy, but take wheatgrass and see the difference,' she says.

Amaranth is
fine, light, gritty, spongy, sticky, herbaceous, malty

It goes with
corn, black beans, chilli, milk, honey, apple, coconut, chocolate

Amaranth

Drive through the countryside of the western United States, especially California and Arizona, and you may notice a reddish weed growing unobtrusively by the side of the road. Look more closely, however, and you will find proof that these weeds are merely displaced or misunderstood plants. This is amaranth, a revered leaf-cum-grain of the Aztecs, a pretty annual herb with red-crimson flowers whose name derives from the Greek for immortal or never-fading. Immortal? Absolutely. Amaranth may be down in terms of public awareness and consumption, but it is by no means out. It's simply sitting there by the side of the road – and in an increasing number of organic food shops – waiting to be rediscovered.

Seven thousand years ago amaranth seeds were a staple food of the Americas, eaten in California, Arizona and Mexico as well as South American countries such as Bolivia, Peru and Ecuador. Its journey to social exile began when the Spanish Conquistadors journeyed to the New World. As the Catholic priests observed, amaranth played a key role in the local religious rituals. The popped grains were mixed with honey and human blood, gleaned from sacrifices, then shaped into figurines and eaten ceremoniously. The Spaniards' revulsion was understandable: in modern parlance, they were grossed out, as most people would be today by anyone eating human blood.

Unfortunately, however, their intrusive response was to place a complete ban on the grain. The *Cambridge World History of Food* speculates that the Spaniards' prohibition of amaranth may have contributed to widespread malnutrition in the sixteenth century. Yet consumption continued in remote areas of Mexico and eventually, in the twentieth century, the Church was forced to acknowledge the grain's popularity, and amaranth was incorporated into Christian rituals, including the manufacture of rosaries.

Another thing the invaders did not like was the way in which locals would go down on all fours to joyously graze any emerging amaranth greens that were still attached to the earth. Ecologist and Arizona resident Gary Paul Nabhan, director of the Center for Sustainable Environments at Northern Arizona University, spent a year aiming to live exclusively from local traditional foods such as amaranth. In *Coming Home to Eat*, his inspirational diary of the project, he says that this grazing prompted the missionaries to condemn the Indians as little more than wild animals. 'It did not matter that their own predecessors – the earliest monastics in the Middle East – also foraged on their hands and knees, delighted by the delicious wild greens their Lord blessed them with,' he says.

Amaranth – and mesquite – once 'the most abundant, widely eaten wild foods during the summer months', also suffered from the misplaced elitism of farmers and ranchers. 'It simply became less and less in vogue for the *gente de razón*, the "civilized residents" of the desert Southwest, to eat wild foods,' says Gary Paul Nabhan. 'As some farmers and ranchers grew wealthy, they used as a sign of their sophistication the fact that their own diets were based on something other than what sustained their livestock.'

Though a different species, amaranth is remarkably similar to another South American seed-grain, quinoa (see pages 62–65), which was also discouraged by European invaders. It has the same flying saucer shape with an equatorial ring (said to be where the nutrients reside) that uncurls during cooking, and offers comparable health benefits to quinoa. However amaranth is much smaller than quinoa, has a milder flavour, and turns sticky (if not downright gluey) when simmered in liquid, arguably making it less versatile in the kitchen. The grains can be cooked whole into a gruel, or boiled to produce tea.

Grinding them for use in flatbreads, cakes and drinks is typical; so too is popping the grains, in which case they are made into sweets, snacks and are also becoming a popular addition to breakfast cereals. In Peru, fermented amaranth is used to make a beer known as chicha. Its vivid leaves can be boiled or fried as a vegetable and, as we have seen, they can simply be eaten raw. Amaranth flowers are also employed as a food colouring, cosmetic rouge, and herbal remedy.

There are several types of amaranth. Those best for grain production are *A. hypochondriacus* (known as 'prince's feather' in England, where it is admired as an ornamental plant) and *A. cruentus* (widely used in Africa, too, as a source of greens). The variety that grows best in the Andes, *A. caudatus*, is commonly known as 'love-lies-bleeding' in the USA. Amaranth is disease-resistant, very easy to grow, and does well in areas afflicted by intense heat, poor soil and drought – it grows quickly in bright sunlight and conserves water by partially closing its leaf pores. However it can also flourish in wet tropical lowlands and mountainous regions. With 50–60 varieties, one can always be found to suit.

The tiny, sand-like seeds – there are around 40,000–500,000 per plant – have an impressive amino acid profile including lysine and methionine, making them a convenient if largely under-utilised vegetarian source of protein. Amaranth grains boast twice the calcium content of milk, thrice the fibre of wheat, and the leaves are an extremely good source of iron – better, in fact, than spinach. Consumption of amaranth is also thought to actively lower cholesterol thanks to its tocotrienols, a type of vitamin E. When cooked it is highly digestible and therefore an especially good source of nourishment for the ill. Whatever its history, it seems the only thing likely to hamper amaranth consumption in future is our lack of familiarity with it.

Buckwheat is
rich, hearty, meaty, pungent, earthy, dark, angular

It goes with
poultry, fish, shellfish, onions, mushrooms, cream, maple syrup

Buckwheat

Buckwheat is the anchovy of the grain world. You either love it or hate it. People who hate buckwheat can't understand how others can bear it and can spot its use at fifty paces. On the other hand, people who love buckwheat tend to want it in everything and can't figure out what everyone else's problem is. Buckwheat appreciation is a rite of passage, a tunnel you go through way after the Channel of Olives, far beyond the Caverns of Marzipan or Goat's Cheese. Once on the other side, if anyone serves a pancake made from bland, wussy wheat flour you tend to think: 'Hmmm… where's the buckwheat?' One day a scientific person may prove it is addictive.

The first thing to know is that, despite the name, it's not wheat and not even related to it. The common English term derives from the Dutch word *bochweit* (beech wheat), a reference to the fact that the plant's triangular seeds resemble beech nuts. The Germans make this association too, in their use of the word *buchweizen*. Buckwheat has also been called Saracen's wheat or Saracen's corn in English, *sarrasin* or *blé noir* in French, and *saraceno* in Italian. These names pay tribute to the role of the Moors in the grain's spread through Southern Europe, but are of bright red herring because buckwheat is not a staple of Moorish cuisine, simply something the Moors traded. Native to Manchuria and Siberia, buckwheat was cultivated in very early times by the Chinese and Japanese and is most likely to have reached Europe via Turkey and Russia in the fourteenth and fifteenth centuries.

Yet most confusing is the misuse of the Russian term kasha. Many stores sell buckwheat – especially the preroasted variety – labelled as kasha, and in many books dishes made from simply cooked buckwheat groats are called kasha. However, strictly speaking, any grain dish simmered in water, stock or milk, whether made with kernels or ground meal, sweet or savoury, can be called kasha in Russia – a key point to remember when reading old cookbooks. Kasha came to be associated with buckwheat in particular only because the grain was so very common in that region.

Buckwheat (*Fagopyrum esculentum*) is not a cereal but one of the leafy seed grains. The herbaceous plant belongs to the Polygonaceae family that includes rhubarb and sorrel, and features attractive white or pink flowers that prompt bees to produce a dark, distinctive, strongly flavoured honey. Like many of the world's less popular grains, it is easy to grow in poor soil and difficult climates, and suits the cold particularly. Other growing advantages are that it is pest-resistant, requires a minimum of fertilizer and is quick to mature. Depending on the variety and area, buckwheat takes only 60 to 120 days to grow, allowing some farmers to produce two crops in a season, or to plant buckwheat in compensation should another grain fail early.

From noodles to pancakes

The earliest archaeological signs of its cultivation are from Japanese sites dated 3,500–5,000BC. While this beats Chinese reports of buckwheat cultivation by a good 2,500 years, it is still thought that the Chinese introduced it to Japan. In both cases they were quick to grind it for use in noodle dough rather than serve it whole as the Russians and other Eastern Europeans do. Indeed, there is no better illustration of buckwheat's vicissitude: the refined, elegant, slippery soba noodle dishes of Japan versus the rib-stickiness of, say, a Jewish casserole of fatty beef brisket, marrow bones, potatoes, onion and buckwheat kernels. Or a Ukrainian buckwheat and mushroom stuffing for roast veal served with a rich gravy of vegetables and stock.

The Orientals are not alone in using buckwheat for noodle dough however. Buckwheat goes into pizzocheri pasta in Italy, and pierogi, which are rather like ravioli, in Poland, Slovenia and Yugoslavia – these can be sweet or savoury, and one particularly curious version fills them with a mixture of cooked millet and curd cheese. The Italians also use buckwheat for polenta and sweet cakes, while in France the common use is as pancakes or galettes, usually savoury, though in the south-western Pays de Léon a porridge of buckwheat and dried fruits called *far breton* is a speciality. Noce is a similar combination boiled in a pudding cloth and sliced to serve.

The unifying characteristic is comfort quality. Even in Japan (especially when compared to sushi and sashimi) buckwheat's role is to provide hearty tucker. A bowl of freshly cooked soba noodles, dipped in a thin sauce of soy, mirin and spring onions, produces enthusiastic slurping noises that are considered good manners there. To gain maximum nourishment, the Japanese then drink the starchy cooking water served hot in a jug and mixed with any remaining dipping sauce. This quick soup is a good source of protein. Serving soy sauce with soba is, by the way, a fairly modern usage – only 100 years ago dashi, the bonito-flavoured Japanese stock, was the norm.

Master chef Yoshio Miyazawa of Takasago restaurant in Shinjuku, Tokyo, teaches soba making in addition to running his highly esteemed restaurant. 'Soba noodles are particularly popular in Tokyo and one of the main foods of Edo culture,' he says. 'However, after World War II American soldiers stationed here discouraged home soba noodle making – they thought it was unclean because production was organic and did not use chemicals.'

Most soba houses in Japan use imported buckwheat from China, Canada or Tasmania, but not Miyazawa. Such is his pursuit of excellence that he now farms his own buckwheat in two market garden areas a few hours' drive to the west and north of Tokyo respectively. It is, he says, simply part-and-parcel of his aim 'to make the most delicious

soba', which in his case is always made from 100 per cent buckwheat flour and, because it is real soba, also priced at a premium.

Like most Japanese restaurants, Takasago serves myriad varieties of the same thing. There are hot and cold soba noodle dishes, noodles with or without soup, cha soba made with buckwheat flour and powdered green tea, goma soba made with buckwheat flour and black sesame seeds. Plus soba gaki, a large, fluffy soba noodle dumpling made only from flour and water and served floating in hot starchy cooking water – it's not as unpleasant as it sounds. It won't taste as good when you're back home – but that's not just because holiday food never does, it's because of Chef Miyazawa's particular attention to growing, harvesting and processing his raw material.

The ground is enriched with charcoal to keep the soil clean, and some South American stevia, a plant that makes the soil fertile and naturally sweet. The buckwheat takes three months to grow, is dried naturally to ensure it tastes its best, and harvested by hand. The stalks are beaten with sticks to remove the seeds, then a machine over 100 years old is employed to separate the seeds from any stones and soil using wind-force.

A second cleaning in another, more modern, machine uses friction to remove the seed husks and grades the grain by size into batches ranging from 3mm to 5mm wide. At first Miyazawa tried to grind the buckwheat into flour by hand but it was too tiring so a machine is now used instead. Nevertheless it is one that does not create heat.

Fresh white flour

What is most surprising to anyone who has encountered packets of buckwheat flour in the West is the bright white silkiness of Miyazawa's flour. It is like talcum powder, not the rough speckled grey flour typical of other buckwheat producers. 'That's because it is fresh,' he emphasizes. 'Sometimes the colour is darker if the producers have ground unhulled buckwheat, however it is often simply because it is old. The older buckwheat gets, the browner it becomes.'

When aged it also begins to taste too strong, even for buckwheat fans. Miyazawa says one of the reasons he is able to avoid adding wheat flour to his noodles as many factory producers do, is that the buckwheat flour he uses is absolutely fresh.

This should be reason enough for you to purchase only pale flour, and light olive-coloured buckwheat grains rather than preroasted, reddy-brown 'kasha' buckwheat of indeterminate age. Toasting the kernels yourself takes only a couple of minutes and is very easy to do – simply heat some oil in a saucepan, add the buckwheat and cook, stirring, until it is browned and fragrant. A Jewish method that keeps the grains distinct is to toss them in beaten egg first.

In fact, cooking buckwheat grains takes very little time all round – only 10–20 minutes, depending on how tender you like them. Without wanting to be patronising – once you are well-travelled or mature enough to appreciate the heady richness of its flavour, buckwheat is an excellent addition to the pantry: convenient, versatile and healthy. For vegetarians in particular it is a good source of protein, as well as the fifth taste sensation of meaty savouriness the Japanese call umami.

In eating buckwheat you'll benefit from its high level of rutin – which increases blood circulation and helps prevents heart disease – plus lysine, B-vitamins, calcium, phosphorus and other minerals. Buckwheat is also considered to be good for strengthening the kidneys and balancing blood sugar levels. Ensure that the buckwheat you buy is not overpowering and you may well be bowled over.

Barley is
honeyed, luscious, pearly, fat, chewy, starchy, rugged

It goes with
lamb, duck, mushrooms, root vegetables, broccoli, oregano, apple, beer

Barley

The cold, remote landscape of the Orkney Islands off the coast of the Scottish Highlands, though blessed with some warmth from the gulf stream, has little in common with the dramatic desert panorama of Tunisia (remember *Star Wars*?). The cuisines could not be more different. Compare the heavy breads and rib-sticking soups of one to the romantic, spicy sauces and light couscous dishes of the other. Yet they actually have more in common than you think. Regardless of climatic zone or temperature, both regions are stark and comparatively inhospitable, and both are home to barley.

Orkney is a collection of islands, with a population of 20,000, at the extreme north of Scotland. So remote is it that the largest island in the area is called 'The Mainland' while Scotland is 'Down South'. Here the local barley or bere (pronounced bear, as in teddy) is preferred to any that folk Down South may care to produce, however there is an element of pride that seeds have been exported to Wales and Canada so that bere may be grown there. Without the support of locals and wandering gourmets, the grain would be at risk of dying out altogether.

Artisan baker George Argo, who is in semi-retirement from the family business he has run for over 25 years, says that only a few farmers grow bere nowadays. 'It does not give a high enough yield to be considered productive, but my uncle has saved the seeds and with some other farmers has kept it going. Bere has more seeds on the ear than the standard varieties of barley. The stalks are longer so they're more prone to fall over, and it takes longer to mature. But it has been grown in Orkney for centuries and it would be a shame if it disappeared.'

George only uses bere meal that has been grown and milled in at Barony Mills in Birsay, Orkney's only remaining working mill. 'There has been a mill on this site since 1873 but there are records of older mills in the vicinity,' he says. The meal traditionally made from bere is significantly different from regular barley meal. Before the stonegrinding process gets underway, the grains are dried in a kiln fuelled by their own husks. 'It makes a tremendous difference to the flavour,' says George, who always works with bere meal and was unimpressed the time he baked with regular barley flour. 'The gourmet will easily tell the difference between bread made from barley and that made from bere.'

The bere meal is typically made into bannocks, which are like a cross between a scone and a soda bread and the traditional teatime staple of the islands. 'We mostly eat bannocks with a bit of butter or cheese,' says George. 'They are served at Orkney weddings at the end of the evening, after the dancing, when we have a second supper. Out come the soups, sandwiches, bere bannocks and cheese.' But modern eating habits and the steady demise of the high tea mean bannocks are not eaten as frequently as they once were. As part of his campaign to keep bere alive, George has developed new biscuits and breads from it and promotes them over the Internet – one admirable example of modern technology helping to save traditional ingredients rather than wipe them out.

In order to achieve the aeration required to produce a deliciously textured loaf of bread, it is necessary to mix bere meal and other barley flours with wheat flour. George chooses a strong Canadian white wheat because it is high in gluten. 'Bere meal is not gluten free, just low in gluten, so it is not that strong,' he says. 'When we set bread aside to prove, the yeast creates carbon dioxide which is trapped by the strands of gluten. If a bread were made entirely from bere meal it would collapse. By mixing it with the high-gluten Canadian flour we are making a dough that can hold itself up.'

Clare Marriage of English organic flour producers Doves Farm recommends making barley bread half-and-half with wheat flour. 'While using barley flour does sacrifice a bit of loaf volume in comparison to pure wheat bread, it has an appealing sweetness.' Indeed, she reckons all barley is due for a revival. 'Dishes such as lamb and barley stew are wonderful and I believe it could be on its way back.'

A classic revival

Those who love barley really can't understand why it has traditionally been classed as a lesser grain. Even in Britain, where barley grows well, it is regarded as a fringe Celtic curiosity, yet its flavour is utterly delicious. Elizabeth David, writing in *English Bread and Yeast Cookery* in 1977, said: 'Those who acquire a taste for it are likely to become addicts. I am one.'

Among the interesting uses for barley that she lists are a pastry crust on Christmas goose pies in Cumberland, and various breads of Devon, Cornwall, Wales, and of the Isle of Man, where the flour is pan-roasted before mixing. She was also inspired by a quote from *Poems Written at Teignmouth* by the English poet John Keats. He was positively lyrical about barley, referring to a romantic place: 'Where close by the stream/You may have your cream/All spread upon barley bread.'

Barley can grow almost anywhere, and was one of the first domesticated cereals. Periodically, various ancient cultures have even favoured it, though some scientists argue that its superiority as a grain for brewing was the most important driver in its spread through the burgeoning civilized world.

Archestratus's *Gastronomy* suggests otherwise. In it barley is classed as one of the gifts of fair-haired Demeter, with some varieties preferred to others. 'The best one can get, the finest of all, cleanly hulled from good ripe barley-ears, is from the sea-washed breast of famous Eresus in Lesbos, whiter than airborne snow. If the gods eat pearl barley, this is where Hermes goes shopping for it.'

Corn is
sweet, milky, juicy, light, refreshing, tender, vegetal

It goes with
beef, cheese, squash, beans, tomatoes, chilli, coriander

Corn
or maize

Italian chef Giorgio Locatelli remembers his childhood Tuesdays fondly. Whereas other families spent relaxing Sunday lunches together, his worked hard at their restaurant, ensuring others enjoyed the Sabbath. Their family's day off was Tuesday, when the children were allowed out of school early, and in winter his father headed down to a hut at the end of the garden to make polenta, a thick corn (maize) mush characteristic of northern Italian cooking.

Polenta is best cooked over a wood fire, so that the mixture takes on a slight smoky pungency – the difference is not unlike that between pizza cooked in a wood-fired oven and a modern oven. The Locatelli's *paiolo* pan was copper inside and iron outside, blackened from being hung over the fire. The polenta was turned with a *canella*, a very long wooden spoon shaped like the base of the pan, cooked and cooked until the mixture was so thick the canella could stand up in it.

When the solid mound of creamy polenta was finally brought to the table it came with a knife and a bowl of water to dip and cut, dip and cut. To accompany there might be fried fish, or rabbit, or a wedge of gorgonzola cheese, but no bread. A meal of polenta is one of those rare instances when Italians feel there is no need for bread.

'Polenta is a great dish for parties,' says Giorgio. 'As a kid I would take a piece of gorgonzola and roll the polenta in a ball around it, then suck out the cheese. The polenta was just warm enough to touch but hot enough to melt the cheese.'

In Italy, Giorgio and family would be called *polentoni* – people who eat polenta – which is as much an insult as a statement of fact. Consumption is very much restricted to the north, where the maize or corn plant flourished once introduced from the Americas, and is typically associated with peasant classes. Although polenta is now associated almost entirely with corn, especially but not exclusively yellow corn, it was formerly made with millet, buckwheat and chestnut flours, and possibly barley in ancient times.

Not everyone takes to it. 'During World War II, polenta was fed to the English army, many of whom developed health problems,' says Giorgio. 'I've met many people who were in Italy during the war and swear they will never eat it again.'

Corn or maize (*Zea mays*) seems to have inspired mixed reactions ever since explorer Christopher Columbus brought it from the Americas to Europe. In 1492 he observed while in Cuba that corn was used there as frequently as Europeans used wheat, so on his return to Europe he took samples to Spain. It was initially well received and subsequent expeditions brought back more strains. However Spain's Charles V decided it was a heathen's grain unfit for Christians. In those days royalty were society's key opinion formers, so in Spain corn became a crop primarily for feeding animals. Today it is rarely used in Spanish cooking, though large-kernelled varieties do go into the manufacture of crunchy cocktail snacks.

The French too have for many years been in denial regarding corn. It is certainly associated with poor farming communities but can be found in traditional baking of Gascony, Bordeaux, Dordogne, Limousin and Auvergne. There, as elsewhere in Europe (and in Africa), corn tended to replace millet in cooking. Regional specialities include *cruchades*, thick corn pancakes, and *canalles*, tall baby brioches with an open texture and rich brown caramel coating.

By contrast, in Romania, corn was adopted with passion by poor and wealthy alike. There a polenta-type dish called *mamaliga* was the daily staple for many years, though it's less popular today. Like the Italians, Romanians use their thick cornmeal mush as a dish in itself, served with milk, cheese and butter, and as a base for other recipes, such as

balmus, baked balls of cheesy cornmeal stuffed with egg, ham and mushrooms. As in Italy, it is cut into slices and fried as an accompaniment to savoury meals. These hot slices may also be simply served with sour cream, yogurt, fried eggs, bacon or onion.

Romanians also like cornmeal mush served cold, an indicator of the esteem in which it is held. Ironically, they view oatmeal the way others view corn: with disdain. Eastern European food expert Lesley Chamberlain writes: 'Oatmeal bread is considered a sign of famine-induced desperation in Romania and oatmeal porridge is unknown. Northern travellers who would not have minded a bowl of porridge from the Highlands loathed their plates of *mamaliga*.'

Corn's association with poverty in Europe was exacerbated when in the early nineteenth century the British encouraged its distribution in Ireland to offset poor grain harvests and potato crop failures. There followed in 1845 an outbreak of potato blight leading to famine, and corn distribution was encouraged again.

Irish food historian Regina Sexton reveals: 'In [the] pre-Famine period, the distribution of inferior quality maize, which was unfit for human consumption, caused widespread illness and soon the corn became a loathsome foodstuff.' Once the famine was established in 1846 people were desperate to eat and rioted in order to get as much as they could.

'Nonetheless, ignorance of the proper preparation techniques was still rampant,' says Sexton. 'This was made worse by the fact that much of the corn was badly ground and was sold in coarse and lumpy form. In many a hovel cauldrons of boiling maize spat ferociously, and young children were whipped from around the hearth in the belief the hissing mass would erupt and scald them.'

It took a while for the Irish to figure out how to use corn (the British would have been no help, they didn't like it), however it was eventually incorporated into

quick breads, dumplings and porridge. The British 'success' with corn as a source of food aid also led them to promote its distribution in Africa.

Corn had first been taken to East Africa by Portuguese traders around three hundred years earlier. Jan van Riebeeck, the Dutch commander who established fruit and vegetable gardens in South Africa, attempted to grow corn he received from the Netherlands in 1655, but this was unsuccessful. It was trekking farmers and Arab traders who dispersed the grain across the continent, and British settlers who first successfully cultivated corn in South Africa after 1820. Indigenous African peoples for the most part adopted it enthusiastically, phasing out other grains such as sorghum and millet, and some tubers, in favour of corn.

African pap

A good cornmeal porridge or pap, says South African food writer, historian and restaurateur Peter Veldsman, 'is to the rural Eastern Cape community what puff pastry or jars of preserves are to the farmer's wife'. In that society it is considered every young girl's duty to know how fine or how coarse to grind the meal. 'And a good pap can only be made from freshly ground meal. A connoisseur will immediately taste the difference between shop-bought and freshly ground meal.'

Chef Johan Odendaal, Veldsman's partner in Emily's restaurant, South Africa's leading venue for Cape cuisine, hails from Free State, where white cornmeal is a staple accompaniment to chicken stew for white people. He has found that many black South Africans express surprise on hearing that white people favour corn, as in many areas they would be viewed instead as potato or rice eaters. All three – cornmeal, rice and mashed potatoes – are offered as standard accompaniments in takeaway chains selling chicken and gravy.

'At barbecues we have a big black pot of pap cooked to medium texture so that it is lumping together, halfway between porridge and couscous,' he says. 'We then stir in corn kernels so that there are yellow specks in the white mixture.'

White cornmeal is favoured in most African communities. 'We have a saying that in bad times we eat yellow porridge and brown bread and in good times we eat white porridge and white bread,' says Odendaal. 'Yellow corn was considered a bad harvest, so we were somewhat surprised by the fashion a few years ago for Italy's yellow polenta.' There was also a period when stoneground corn was considered primitive, 'but we've gone past that and are now longing again for stoneground meal.'

An Eastern staple

Somewhat unsung is the use of corn in Asia, where it was also introduced by Portuguese traders. We've all seen baby corn in vegetable stir-fries. In India the largely vegetarian and culinarily experimental population quickly saw corn's potential in flatbreads, dumplings, curries and snacks. China employs it in combination with other grains, for example cracked and cooked with rice, or ground into noodles. A fine sweet dim sum or tearoom snack features corn custard inside steamed puffy balls of yeasted wheat flour dough. Corn is especially important in Indonesia, where it supplements the rice crop. Seventy-five per cent of the corn grown there is consumed by humans and it supports some 18 million people. Bijon noodles, made from fermented corn mash, are one of the ways corn is used in Indonesian and Malay cooking.

In South Africa, as well as Egypt, Europe and the Southwestern United States, corn has been linked to pellagra, a niacin-deficiency disease, and kwashiorkor, a protein-deficiency disease. These illnesses are not caused by corn but by extreme poverty, which as we have seen has also been associated with corn consumption. As far back as the early 1900s, American epidemiologist Joseph Goldberger concluded that pellagra was caused not by eating corn but by the financial inability to obtain complementary foods to eat with it. He and other physicians recommended changing agricultural and dietary habits to include a wider variety of foods, which saw the elimination of the disease.

But the damage was done. Although corn as a garden and field plant had established itself in regions of Europe during good times, when bad times came generations later and the poor had nothing else to eat, corn was labelled the source of malnutrition and has had an image crisis ever since. We tend not to hear that it is rich in complex carbohydrates, essential fatty acids, potassium and magnesium, only that it is (like other cereals) low in amino acids.

On home ground

In its native land, the Americas, however, corn is considered a divine gift and the stuff of which people are made – parallels are often drawn between the corn plant and the human body, and between the plant's growth and human life stages. Translations of native American words for corn include 'Our Mother' and 'She Who Sustains Us'. It is used in all manner of dishes and drinks, as well as in religious and social ritual.

Even America's European immigrants willingly adopted corn into their diets, partly through necessity but also perhaps demonstrating their enthusiasm to settle in the new land and make it home. The Dutch adored the native Indian dish *sappaen*, a cornmeal mush they seemingly appreciated thanks to their love of porridge in the Old Country. It remains a part of traditional Dutch-American celebrations, served as the first course of a banquet and eaten from a large communal bowl placed in the centre of the table.

Benjamin Franklin, American statesman, scientist and co-author of the *Declaration of Independence*, was corn's most prestigious exponent. In 1766, responding to criticism that corn was un-nutritious and fit only for swine, he wrote: 'Indian corn, take it for all in all, is one of the most agreeable and wholesome grains in the world – johnny and hoecake hot from the fire is better than a Yorkshire muffin.'

The emerging United States' love affair with corn began very early. In the spring after the *Mayflower* landed in 1620, friendly natives of the Wampanoag tribe taught the fledgling, starving Plymouth colony how to grow the crop, and its successful harvest was the basis of the first Thanksgiving. A key part of the natives' instruction was their system of farming which saw corn, beans and squash – a perfectly balanced meal – grown together, then cooked together. Succotash, a classic dish of corn and lima beans still popular today, is said to have been served at the first Thanksgiving – as was popcorn.

Also important, particularly in Latin America, is the preparation of corn with a calcium hydroxide solution. The process, which involves soaking and cooking the corn with wood ash, seashells or crushed limestone, is known as alkali processing or nixtamalization. It loosens the grain's outer hull, and makes its niacin and protein content more available to the body and the kernels easier to digest. This elemental technique did not make it to Europe with the corn plant, and this has undoubtedly played a role in the perception of corn as an un-nutritious food, yet is fundamental to native American diets and features in corn products including tortillas, hominy and posole.

It was the sheer ubiquity of *Zea mays* in the Americas that led to it being named corn. At one time the word 'corn' referred to any grain that was the staple of a particular region, but maize was so common that the generic word eventually became the specific and an alternate name of Virginia wheat was phased out. Today the corn crop of the USA is as large as the country's combined production of other grains, although much of it is for animal feed and non-food uses. Navigating the myriad varieties available and products made from them can be confusing, but exploration is worthwhile.

Corn differs from other cereal grasses in that it produces a multi-rowed ear of kernels attached to a cob, the whole ear, rather than each individual grain, enclosed in silks and husk. The cobs or seed heads are distinctively large, in some cases as big as 60cm/24in, however this was not always the case and tiny cobs of only 2cm/¾in have been found on archaeological digs in Mexico. Separate male and female flowers are borne on the same plant, but they are set far apart. Human intervention is necessary in reproduction because the enclosed kernels do not fall to the ground to reseed themselves.

Understanding corn

There are five families of corn, each with hundreds of varieties in several colours including white, yellow, orange, red, brown, blue, purple, black and speckled. Sweetcorn is the type grown specifically for consumption as a young vegetable, though in the Americas it is also sun-dried then rehydrated. Baby corn, the miniature version popular in Oriental and Southeast Asian cooking, is not simply corn picked early, but specially bred to be eaten at that stage.

Popcorn is a very hard variety, with a tough hull and endosperm. The moisture inside the grain turns to steam when heated, and the pressure builds, causing it to explode. It is not possible to make a dish of popcorn from just any old corn (though it is possible to pop some other grains).

Dent corn is the most common variety grown in the USA and is used in food manufacturing and as animal feed. The name refers to the dent that develops in the top of each kernel as it dries. Dent corn is sometimes made into posole or hominy, alkali-processed grains of corn. This in turn is freshly ground (or wet-ground) to give *masa*, the dough used in tortillas and tamales, which may also be made from *masa harina*, or dried posole meal.

Flint corn is used to make posole and its products too, and is the variety used to make the coarse stoneground meal for polenta. Curiously, given its primary use in Europe, it is comparatively difficult to grind as the small grains have a hard skin, however it has a good flavour.

Flour corn is a softer variety that is used for manufacturing cornflour, fine cornmeal and pinole, a type of wholegrain corn flour used to make drinks in Mexico and Southwestern USA. This is the family to which the distinctive Hopi blue corn belongs.

Alongside these native varieties and regional dishes, corn is big business and its image as a wholesome grain of the Americas has been somewhat tarnished by over-production, monoculture rather than sustainable farming, and an association with genetic engineering. It has been suggested that technological developments in corn agriculture, rather than benefiting disadvantaged farming communities as marketing professionals would have us believe, in fact makes them more vulnerable.

Corn's reputation as superlative animal feed comes from its high-calorie, high-fat, low-fibre content that fattens stock and poultry quickly and at a lower cost than other grains. However, some argue that animals fed a diet of corn produce meat that is higher in unwelcome fats than that of animals grazed traditionally on grass fields or ranges. Yet it is anticipated that demand for corn will increase in line with increased consumption of animal foods (fed on corn) and rising demand for processed snack and convenience foods. Food lovers must put their money where their mouth is and seek out corn products of estimable provenance.

Wild rice is
spiky, nutty, chewy, elegant, earthy, meaty, fragrant

It goes with
mushrooms, celery, poultry, fish, nuts, cream, orange, parsley

Wild rice

What's wrong with manoomin or tuscarora? These names for the species *Zizania palustris* are more accurate than the term wild rice, which presents food writers with the task of explaining, no, it's not rice, and it's usually not wild either. Indeed, this North American water grass today carries its common moniker merely because early fur traders in the United States and Canada likened it to the 'real' rice, *Oryza sativa* (see page 56), and subsequently misnamed it. Wild rice has also in the past been termed wild or 'false' oats by early immigrants, but in using the word manoomin, translated as 'good berry', the native Americans are actually likening it to a fruit. Not so long ago the scientists couldn't make up their minds either – which is why some reference works use the term *Zizania aquatica* to describe wild rice, though officially that now refers only to a non-edible species of the same genus.

Confused? You are not the only one, for even in these culinarily sophisticated times, some restaurateurs and grocers think any long black rice, such as Thai sticky rice, should be called 'wild rice', leading to misunderstandings all round. Pedantry may seem excessive but is important when an increasing number of people suffer from food allergies and should not be forced to avoid any more foods than absolutely necessary. And no one would want to avoid wild rice, for it is one of the most delicious and visually enticing grains.

There are a few Native Americans such as the Ojibwe of Saskatchewan and Minnesota, who continue to collect wild rice from the wild, paddling out in small canoes during late August and early September and smacking or beating the water grasses to extract the grains. This period is known as *manoominike-giizis*, meaning the wild rice moon. For some harvesting it is a spiritually important ritual, for others a means of acquiring a personal supply and enjoying a pleasant day out with the family.

Not very long ago wild rice was the base of many small businesses but these swiftly declined in number when the technology was developed to cultivate it on a large scale. Now the majority of 'wild rice' on sale is farmed (it's sometimes called paddy-grown or cultivated wild rice on the packet), and therefore considerably cheaper than the truly wild, hand-collected grains. Nevertheless, much of the cultivated wild rice available, especially that from Canada, is grown to organic standards.

Ancient traditions

Native Americans consider wild rice to be the Gift from the Great Spirit, a sacred food, and treat it with honour. Archaeological evidence indicates that wild rice has been an important native food for at least 1,000 years, though the Ojibwe have been harvesting it only for the past 300 or so. The ancient belief system suggests that wild rice was a gift because it was not sown by people, it simply appeared in the water when the spirits wanted to bestow it. In addition to its consumption at ceremonies, wild rice was used as medicine and in everyday meals.

At the customary First Rice offering, the elder Ojibwe women were (and still are) particular that perfect, unbroken grains be prepared. A traditional dish on these occasions was made by popping the fresh, soft, green grains in bear fat or deer tallow. Incidentally, fresh wild rice doesn't explode in the noisy fashion of popcorn, but rather puffs out slowly to give a pillow shape. Popped wild rice was popular on other occasions too, such as during sugar camps, when it would be covered in maple syrup, and on winter travels, when it was a convenient snack food, crushed and shaped into cakes with deer fat, melted sugar and dried berries.

The traditional method of harvesting is a model of sense and sustainability. Reconnaissance missions were made in the weeks leading up to wild rice moon and people – traditionally women – would mark out their allotment by tying the stems of rice together. This had the advantages of making the grain easier to collect and preventing damage or loss of grains from birds and strong winds.

Harvesting was not permitted until the ricing chief officially opened the season. Collecting the grains was done in canoes, ideal for moving through tall grasses in shallow water. A person would stand at the back pushing the boat by pole, while another sat at the front with two sticks. One stick was used to hold the rice stems over the canoe, the other to gently knock the heads of the rice stems so that the ripe grains fell into the canoe. Being light green, the stems of wild rice were easy to distinguish from the surrounding cat's tails and bulrushes. Using a light touch to knock the heads meant the unripe grains stayed on the stems to be collected another day when the team would make a return trip, being sure to move the canoe through the channels made on earlier expeditions that season. (It was this gradual maturation of individual seeds that was to make the introduction of mechanical harvesting particularly challenging.) Any grains that fell into the water rather than the boat sank to the bottom of the lake to seed, to appear in future years as a gift of the spirits.

Once the grains were collected, they had to be dried or cured immediately, or else kept wet until it was convenient to dry them. Sometimes the heat of a fire was employed to assist the natural drying qualities of the air and sun. The grains would begin to darken and harden while curing. Then they would be parched, effectively a drying and roasting process which lowers the total water content to six or eight per cent – that's around 40 per cent less than the moisture content of the freshly harvested grains, so that it becomes quite a different product. The parching, done in a metal cauldron or galvanized washtub over a slow-burning fire of poplar, birch or tamarack, also destroyed the germ of the grain so that it could not sprout and could therefore be kept for long periods.

To hull the parched grain, a pit would be dug in the earth and lined with wood, clay or even cement, then a blanket or other material. Methods differed: some people would put the grains in a skin bag and then place that in the pit. The rice was then 'treaded' or 'jigged', usually by a man wearing new or very clean moccasins. He would carefully walk on the grains, moving his feet in a circular motion and using poles to suspend himself just enough so that adequate pressure was applied for the hulls to come away without breaking the grains. This was one of the first processes of wild rice production to be mechanized and today it is highly unusual even for those adhering to the traditional collection methods to process their own grains.

Cultivation of wild rice

The breakdown of traditional wild rice harvesting probably began in the mid to late nineteenth century, when the custom of waiting for the ricing chief to declare the season open was not strictly observed. Some Ojibwe believe this demonstrated a lack of respect for the crop. It led to a perception that wild rice was a natural asset to be exploited, and could be collected using motor boats and mechanical harvesters which soon decimated the beds. In Minnesota, steps were taken to correct this problem with the introduction in 1939 of a Wild Rice Act, restricting harvesting to people with a licence, and demanding the use of traditional methods of collection. Binding the stems is no longer done, however it is once again necessary to wait for the official opening of the season.

Cultivation of wild rice is not a new idea, nor one restricted to Caucasian people. Ojibwe legends tell that after a young man, Wenabozhoo, discovered

wild rice, women from his tribe planted seeds in the lake near their wigwams in the hope that it would grow there. Re-sowing some of the first grain of each season became a practical spiritual tradition that continues to this day when wild rice collectors donate a small portion of what they have harvested to replenish the natural rice beds.

At an early stage in the colonization of North America, wild rice cultivation was encouraged as a means of assisting game hunters. Both Native Americans and whites took advantage of wild fowls' attraction to wild rice, the latter shooting the ducks and teals, the former catching them in the water in nets as the birds dived for the precious seeds. But the benefits of cultivating wild rice as a food crop were also appreciated and eventually, in 1917, the first mechanical harvesting took place. The first cultivated wild rice paddy was established in Minnesota in the 1950s, initially hand-harvested using very similar methods to the native collectors, but soon harvested mechanically. This became possible once agricultural scientists recognized that – as with wheat and corn – they could select from a crop individual seeds with favourable characteristics (in this case late-maturing grains) that would then be replanted, ultimately allowing a paddy of wild rice to be harvested by machine in a single pass, rather than by hand over a period of several days.

The development of the industry has made wild rice cheaper and more readily available, but some Ojibwe believe it to be the ultimate desecration of a spiritually significant grain that played a central role in tribal life. Native Americans maintain that the truly wild, wild rice of the various northern lakes boast significantly different flavours, and that these can be appreciated as a connoisseur does wine. Even today, these people trade their locally harvested grains for those of other tribes in order to enjoy fully their Gift from the Great Spirit.

Millet is
mild, beady, creamy, light, golden, sweet, alkaline

It goes with
beans, eggs, milk, cumin, parsley, green vegetables, berries

Millet
teff and sorghum

If I wanted to eat millet, I'd sit on a perch and tweet. Certainly that's the view of many who have yet to realize its delicious versatility or impressive nutritional qualities. Yes, millet is great bird food. It increases the egg yield of chickens and is reputably capable of restoring the voices of mute canaries – but this is far from anyone's thoughts when enjoying millet's use as a star dessert at the leading modern Japanese restaurant Kozue in Shinjuku, Tokyo. There the tiny beads are delicately bound with beaten egg, shaped into little dumplings filled with a few morsels of sweet aduki beans and lightly fried before being served floating on an intense red sauce.

This is, of course, dining at its most refined and elegant, special occasion food, the food of the very wealthy. However, throughout the rest of Asia, as well as in Africa and Europe, millet is more usually seen as the grain of the poor. 'Millet we use often,' admitted one highly respected Indian chef, but when pressed for more information he was dismissive: 'I don't want to tell you about it. It's poor man's food, not to the English palate.'

The Chinese classify millet (along with wheat, rice, barley and soya) as one of their 'five sacred crops', however their common name for it, *xiaomi*, means 'little' or 'lesser rice'. Consumption of millet as a staple food is diminishing even in the Northern areas where it is widely cultivated. Recipes employing millet tend not to be published in Chinese, nor are you likely to find it on the menu of your local takeaway. Ironically, millet scores better in nutritional analysis than rice or wheat, and despite accusations of blandness, it beats white rice hands down on flavour when lightly toasted before boiling. What respect the grain still garners in China is earned for its role as animal feed. The Chinese consider millet and other grains to be life-giving because they give life to the animals that produce the meat part of their diet. It's also used in distilling.

In Europe millet was also once a vital crop but its use has largely been phased out as robust varieties of other grains have been introduced. Many country dishes and baked goods now made from cornmeal, rice or wheat were formerly cooked with millet – polenta is one of them. Now it seems to be only in times of crop failure, war or famine that millet is fully appreciated. It grows so quickly that farmers can plant it late in the season, after other more popular crops have failed, and it will mature in time for regular harvest. Kept dry and in the dark, millet grains can be stored for up to 20 years without attracting insect damage. It also hydrates well, so that a small amount of millet will feed more people than the same quantity of rice.

There are several varieties of millet, however the one typically sold in the West is proso (*Panicum miliaceum*). If you want to sprout millet you must buy grain marked 'unhulled'; it looks like it is wearing a shiny plastic coat. Millet flour can be delicious – a great addition to cakes and breads – but it turns rancid fairly quickly and most of what is on sale is past its best, leading to inevitable disappointment on cooking and eating. Therefore only buy millet flour from a store with a very high turnover, and buy in small quantities and often. If you want to use millet flour regularly, consider grinding your own – it's easy to do using the grinding attachment of most modern food processors. Ready-puffed millet for making breakfast cereals and sweets is available in many health food shops.

Those who have had the good fortune to travel to areas of sorghum cultivation tend to be rather perplexed that it is not more widely available as food for home use. In South Africa it is malted and ground into maltabella, a wonderful chocolatey-tasting porridge that is a favourite healthy breakfast and the kind of comforting family food that makes expatriates turn misty-eyed while reminiscing about home.

There are several varieties of this grain, a close relative of sugar cane, though *Sorghum bicolor* is the one suited to human consumption. It is widely used throughout Africa, and in the first millennium BC was taken as a foodstuff to India, where it remains a vital crop. Slave traders introduced sorghum to America in the mid-nineteenth century. Today it is grown primarily as livestock feed and for making molasses, however it can be used in a variety of porridges, breads and snacks and also eaten as a fresh vegetable.

Sorghum's continuing popularity in its native Africa is guaranteed by its hardy ability to grow in areas of low rainfall. Unfortunately it also tends to attract birds that can devastate the crop before it is harvested. In areas where this problem is acute, high-tannin varieties of sorghum act as a deterrent but also make the plainly cooked grain less palatable, so locals tend to use it in fermented foods to improve the flavour.

Teff (*Eragrostis*) might be considered the national grain of Ethiopia, were it not that the people of the country have in recent years been actively encouraged to cultivate crops more popular on the international market. It is extraordinarily tiny and difficult to grow, but the Ethiopians hold teff in high regard and it is the essential grain for their national bread injera, a large fermented flatbread that doubles as a plate for spicy stews. Teff is also used in brewing. It is native in particular to the highlands of Ethiopia, however a wild variety of the species is harvested by the Sudanese to supplement their crops of sorghum. Around the world it has limited availability on the retail market to date, but has been grown in the USA.

Ethiopians have told me they would not use teff in cooking except to make injera. People from other countries, meanwhile, report it can be used to make porridges, resulting in a mushy mixture not unlike spongy boiled amaranth – proof that just because something can be done, doesn't mean it should be.

Oats are
unctuous, soft, rich ambrosial, comforting, rugged, toasty

They go with
bacon, game, lamb, nuts, soft cheese, berries, muscovado

Oats

The 1980s were a distinctive era in popular culture. There was *Fame* and *Flashdance*. Lots of leotards and frilly ra-ra skirts, big puffy haircuts and big puffy shoulder pads. Men in make-up, everyone in lace. Duran Duran and A Flock of Seagulls hit the big-time. So too did cholesterol and oat bran.

Throughout the English-speaking world, housewives were feverishly stirring oat bran into cereals, breads, cakes and cookies – anything – in an attempt to combat the high cholesterol levels of their husbands. Cholesterol had recently been identified as a key contributor to heart attacks and oat bran labelled as the best weapon against it. Before long it was legal for manufacturers to claim products with oat bran could prevent heart disease (somewhat of a milestone in labelling regulations), and everywhere you went seemed to be selling American-style high-bran muffins with a revolting dry texture. Yet the public were determined to learn to like them, even if it killed their taste buds.

This food-as-a-pill approach to eating is actually completely unnecessary. True: oat bran is particularly rich in soluble fibre, but there is absolutely no need to separate it from the oat's sweet creamy kernel, which adds so much flavour to dishes that eating it regularly is a pleasure.

Sarabeth Levine, who is New York's queen of breakfast, understands exactly this dual appeal of oats, health-plus-sensuality, exemplified by the four types of hot oatmeal porridge offered on the menu at Sarabeth's Restaurant, as well as the oatmeal pancakes that are served invitingly with strawberries and cinnamon. 'I love oatmeal,' she says, 'and have it in its simplest form when I eat it – maybe with just a little milk. We serve it every day and feel it's an important part of the menu now. People prefer it to other cereals because it's natural and good for you. It's very filling, nourishing and tasty and takes you through the morning.'

Humble beginnings

Oats (*Avena sativa*) are the only whole grain many people ever eat, and have been perceived as healthy since very early times; however it was not until the late twentieth century that large parts of society took this as motivation to consume them. Oats have been disparaged throughout history as being only suitable for animals, yet Mongolian warriors, Crusading knights and Scottish soldiers marched on them.

Such was the low esteem in which oats were held it is somewhat of a miracle the grain made it to Scotland from its place of origin, the Middle East. The Egyptians considered it a weed, the Romans thought it animal fodder, and it was the Bronze-Age Gaulish and Germanic tribes who first put oats on the culinary map. The earliest written record of oat consumption was by the Roman Pliny, who observed that Germans 'made their porridge of nothing else'. Today oats are so inextricably linked to porridge that many don't realize it can be made with anything else.

As the grain moved north to colder climes and acidic soils it improved in quality. The oats took a longer period to grow in Northern Europe, Scandinavia and the British Isles, so the grains grew fatter and juicier. 'Even those grown in the south of England [where the weather is milder] are not nearly so well filled as those in Scotland,' says Scots food expert Catherine Brown.

Oats are also important in the traditional Irish diet, a vital complement to the potato and valued since the Middle Ages for their nutritional value and keeping qualities. They were a favourite of monks and recommended for the sick (and menstruating nuns!), who would eat them in the form of porridges and gruel. Oats increased in importance during the Irish potato famine and flat, dry oatcakes, when properly made, had such a long shelf-life that many of the people who emigrated to America set out with supplies of oatcakes to eat on the long journey.

But it was the Pilgrims who first established oats in the Americas. Their association with religious devotion continues today in the form of the world-famous Quaker brand cereals. Launched in 1877, this was the first US trademarked cereal and the first example of mass-marketing promotion techniques. It seems such a simple act – to put oatmeal in a box with a logo to sell through stores – yet at the time it was thoroughly innovative. Oatmeal had previously only been sold in bulk bags and the new, convenient packaging prompted an increase in consumption.

Oats are one of several grains mentioned in a letter dated 1626 from Pieter Jansen Schagen, deputy to the State General in The Hague. It is the only record of the purchase of the island of Manhattan by the Dutch 'from the wild men for the value of sixty guilders'. It says: 'They sowed all their grain in the middle of May, and harvested it in the middle of August. There being samples of summer grain such as wheat, rye, oats, buckwheat, canary seed, small beans and flax.' By the mid-1600s oats formed just over half the New Netherland crop; less than 43 per cent was wheat, but as wheat drew a higher price it was the most important cash crop, and the vast bulk of the oats were destined for animal feed.

Today around 16 per cent – only 16 per cent – of world oat production is consumed by people, but this figure is steadily rising. Oats are well established in the kitchens and food manufacturers' laboratories of Britain, America, Australia and South Africa, particularly for breakfast foods and baked goods, however they are used in other countries too. They are a little-known staple of northern China and Mongolia. In France they go into oat breads, including *pain d'avoine à la crème*, made with cream, and oatmeal soup, *potage à l'avoine*, a simple mixture of stock, oatmeal, butter and salt. The Czechs make a rich puréed soup of potatoes and oats that is enriched with milk, butter and egg yolks.

Rice is
fragrant, tender, light, delicate, starchy, refreshing, absorbent

It goes with
meats, poultry, seafood, pulses, nuts, herbs, spices, milk

Rice

There was a time when buying rice (*Oryza sativa*) was a simple choice of white or brown, and for most of us that meant white. They are essentially the same product: one stripped bare, leaving little to the imagination, the other puritanically still wearing its seed coat which, as coats go, tends to be tightly-buttoned sackcloth rather than linen or cashmere.

But a bright and sexy new collection of colourfully coated rices is finding its way onto store shelves. Some are pure haute couture – traditional varieties previously cultivated only for aristocracy – while others are newly arrived models. What's notable, however, is that these rices tend to flaunt their technicolor dreamcoats in gourmet retailers and supermarkets rather than health food stores.

Even many Chinese food enthusiasts are unfamiliar with the Imperial green and Emperor black rices now on sale, and their availability in China is limited. Legend has it that they were grown exclusively for royalty, and the black in particular, sometimes known as Forbidden rice, is thought to be beneficial in traditional Chinese medicine, especially in the treatment of diabetes. Flour is also produced from Emperor black rice, though at the time of writing it is not widely available.

In Piedmont, the same Emperor black rice has been crossed with an Italian variety to produce Venere Nero or Black Venus rice. This new version is being championed by top restaurants such as Vissani in Baschi, Umbria, where it is boiled then tossed with fresh asparagus spears and a sauce of onion and garlic.

Rice grower Michele Perinotti of Gli Aironi who produces Venere Nero and a wholegrain red variety (rosso integrale), says the Italian interest in wholegrain rices is growing. Even his artisan 'white' rices are *lavorazione morbida* or 'soft work', meaning a thin layer of seed coat is left on the grain. 'Besides the many nutritious properties (including antioxidants, selenium and fibre), there is a great deal of flavour there,' he says. Milling results in the loss of nearly 50 per cent of the grain's vitamin B and iron content.

While the black and red rices take longer to cook than industrially milled white rice, they are easy to use and are best simply boiled, drained and tossed with sauce as is done with pasta. 'They are mostly served with fish, but also vegetables, and are exceptional when used cold in rice salads,' says Michele, who also grinds his black rice into flour and makes gnocchi, pasta and biscuits from it. An important factor to note is that, unlike with pale brown rice, the colour leeches from brightly coloured rices and can taint other ingredients in a dish when cooked or tossed together.

Discovering red rice

In producing the rosso integrale Michele was inspired by an old red-and-white-striped variety called *Ostigliato di Mantova* that 'has been around forever but was never really appreciated until now'. Stray red rice has for a long time spontaneously appeared in fields and was considered an annoyance to be discarded. 'It wasn't nice to have a few red grains on your plate among the white as it looked like an insect or something horrible,' says Michele. The turnaround for him and other producers came when they realised that they could plant the red rice alone, 'not separately from the white rice but indeed as the main rice in their production', he explains. Thus the 'defect' was turned into an attribute, for which we are now being charged premium prices.

The moment of awakening came to French farmer Serge Griotto when he was walking in his fields one day and noticed one ear of grain significantly taller than the others. He picked it, opened the seed husks and saw that the rice was garnet-red in colour. The few grains he saved and planted became a crop of several hundred kilos within three years.

Camargue red rice has become something of a modern tradition in the Camargue and Provencal regions of France, where chefs tend to serve it with fish and chicken. The Camargue and Gli Aironi red rices retain a firmer texture and more vivid colour than Bhutan's red rice, now being sold in speciality stores and supermarkets, which is typically milled to leave just a thin coat on the rice, thus making it quicker to cook.

'The current fashion for coloured rices is largely the result of the success of Camargue red rice,' says Mark Leatham, its British importer since 1993, adding that it is still early days for the coloured rice market. 'As long as the colours are natural, they bring mystique to a rice dish,' he says. 'The total rice market is growing at over five per cent per year, so there is every chance the coloured rice sector will continue to grow.' Like other importers, he is constantly on the lookout for new rice varieties. 'There is more to discover,' he says.

Happily people's enthusiastic consumption of these intriguingly coloured rices has the related benefit of improving their intake of the fibre and other nutrients contained in the coloured seed coat. Milling results in the loss of nearly 50 per cent of the grain's vitamin B and iron content. Yet wholegrain rice – when brown – has for years been the object of ridicule and disdain. 'Brown rice and sandals' is a commonly used derogatory phrase applied in relation to the health food movement, and card-carrying vegetarians in particular.

People in the English-speaking world are not the only ones to have derided brown rice however. In many Asian cultures white rice is perceived as a status symbol, in part because it is traditionally more expensive than the wholegrain. The Indonesians, for example, realise that milling removes nutritional value and fibre, and according to the Indonesian-born writer and rice expert Sri Owen: 'They are as conscious of the importance of diet and nutrition as Westerners are, but the cultural insistence on whiteness is too strong for them.'

While living in Tokyo, Shirley Booth, author of *Food of Japan*, found it difficult to buy brown rice without making a special trip to a health food store. The Japanese consider it an 'unrefined food for unrefined people' and feed it to prisoners. 'Its image must be bad if it is seen as a punishment,' she says. While living there she developed a liking for *haiga mai*, Japan's version of semi-pearled rice with the husk removed but germ intact. 'It's more nutritious than white and more elegant than brown.'

Reviving brown rice

In Spain, the prestigious Calasparra rice growers of Murcia produce a similar product they call semi-integrale rice (probably my favourite, by the way). They also sell integrale, otherwise known as wholegrain brown rice, however this is intended almost exclusively for export to trendy organic shops in other countries. It generally struggles to find distribution in the cost-conscious traditional health food trade because of its premium price, a figure commanded not just for the rice's organic production but for its rarity and gourmet status.

Calasparra rice is the only rice to be classified under the country's prestigious *Denominaciones de Origen* or DO system, meaning that law regulates the practices and standards of its production. All of the distinctive hand-sewn muslin bags that the rice is packed in are individually numbered to guarantee their provenance and provide traceability.

The region could not look less like the lush green paddy fields of Asia – it is more like Marlboro Country or the set of a spaghetti Western. What pine trees were covering the mountains were destroyed by fire in decimating blazes started by lightning. Rice cultivation in Murcia dates back to the end of the tenth century, when the Arabs introduced the grain and the necessary watering system to produce it before subsequently taking rice on to Italy.

A cooperative of around 160 farmers grows rice, wheat and maize – plus vegetables – using crop rotation in an area totalling around 900 hectares/220 acres. Each worker decides what crops to plant each year; alongside the rice there may be broad beans, broccoli or tomatoes. In the vicinity there are also apricot, almond and olive trees.

The series of individually owned small fields is gently, almost imperceptibly, terraced. Water from the fast-running, trout-filled, pale-olive coloured river is channelled into the fields and provides electricity for all the houses in the region. At source, the water is strong enough to present a challenge to canoeists. In the paddy fields there is a dichotomy: it looks completely still, apart from the occasional ripple, yet is constantly moving, and is so clean that one can drink directly from the paddy.

Wheat cultivation does not require flooding of the small rectangular fields yet demands about as much water as does the rice. Rice grains are generally large, but if they were simply scattered over the ground and then flooded they would float to the surface, so wet sacks are placed over them to keep them down until they germinate and establish roots. The steady flow of water through the area also discourages flies and mosquitoes and there is no tendency towards plague insects in the area, making organic production of the rice relatively straightfor-ward – although not all of the production is organic.

Come the harvest in October, each farmer collects his own grain and takes it to the cooperative where it is weighed, then dried. The humidity level is reduced from 100 per cent to 13–14 per cent to prevent fungus growth. The grains are processed, the bran removed as necessary and fed to animals, and the rice is piled into its trademark white muslin

sacks. Local women use vivid red cotton to hand-stitch the tops of the bags in criss-cross fashion to seal the rice inside, then attach the numbered tags explaining the provenance of the grain.

'Calasparra is unique – the only rice with government quality control – and it's one of the best rices you could use,' says itinerant Spanish chef and paella expert Carlos Vargas. He thinks the wholegrain version is delicious but adds: 'People in Spain would say you were mad if you served them brown rice. We are like Italians in that respect – very traditional – and it is very difficult to introduce new things. However, with the increasing interest in health and vegetarianism, very slowly, recently there has been a slight change.'

Cooking uses

He is keen to stipulate that wholegrain rice should not be used for paella or any other of Spain's vast collection of traditional *arroz* dishes. This is not bias on his part, but a deep understanding of the science of good cooking, and is also true of Italy's risotto. The wholegrain rice takes far longer than pearled rice to cook properly and requires significantly more liquid. This is not a problem if the liquid used is just boiling water and the rice is simply intended to be served as a plain side dish, however for traditional Spanish and Italian rice dishes cooked in an open pan, the cooking liquid is a well-flavoured broth or stock. The layer of bran prevents absorption of the flavoured cooking liquid, while at the same time more and more is being added to cook the grains, resulting in a over-seasoned, unpleasant mixture.

In addition, the ingredients of the stock and any flavouring vegetables in the pan lower the overall boiling temperature. This means that while wholegrain rice might take 40 minutes to cook in plain boiling water, it can easily take 90 minutes in a risotto-style dish and still be a bit too chewy. Who of

us wants to stand next to the stove stirring for an hour and a half at all, let alone for a dish of disappointing flavour?

Par-boiling (half-cooking) the rice in boiling water before starting the dish is one method of producing a wholegrain version of paella or risotto but don't kid yourself that it is authentic, or that the result will be particularly tasty. Being very plump and chewy, whole short-grain rices are not well suited to summer cooking, though passable salads can be made with them. They are best added to wintry stews. 'A nice lamb stew with olives that takes a long time to cook on the stovetop or in the oven works very well,' says Carlos. Vegetarians wanting to make use of whole short-grain rices should match them with dried beans.

Similarly, white long-grain rices such as basmati, should not be used in paella or risotto. 'You need a fat rice to absorb the taste of the other ingredients while also staying firm,' says Carlos. Short-grain rice does not overcook as easily as long-grain rice. 'The process of absorption is slower,' he explains. 'Basmati needs to be cooked fast and therefore does not have as much time to absorb the flavour of the other ingredients.' By the time it has taken on the required level of flavour, the texture of the grain is ruined. The lesson: just don't do it.

As Carlos says: 'Any rice tastes good if you know how to treat it.' Knowing how to treat it involves knowing what type of rice it is. There are many ways of classifying the myriad rice varieties available, and many of the new rices produced around the world do have suitable cooking qualities for traditional dishes. As we have seen, describing rice merely in terms of its colour is redundant. For the cook the key issues are whether or not the rice is milled, part-milled or wholegrain; whether it is a long, medium or short grain; and the rice's degree of stickiness. The rice may also boast a particular aromatic quality – as in Basmati, Pecan rice, Thai fragrant rice, and Pearl

Jasmine rice – however this will not affect the cooking process.

Plain-boiled rice for use as a side dish can be made using long, medium or short-grained rice. Indian and Chinese cuisines favour long-grain rice, while in Japan and some Southeast Asian cuisines sticky rices, usually short-grained, are used. Thai black sticky rice, being wholegrain, is less sticky than its milled equivalent, but is a long-grain variety.

Pilaffs and biryanis are properly made using long-grain rice. Risottos require fat, medium to short-grain rices, although the superfino grades are comparatively long. Arborio, carnaroli and vialone nano are the best-known Italian risotto rices.

Short-grained rices are best for puddings, particularly slow-cooked ones, where their fat starchiness enhances the texture of the dish. They are sometimes sold as pudding rice. Sticky short-grain rice is essential for sushi.

Food for the world

Rice is grown in 112 countries and on every continent except the Antarctic. From its origination in India and first cultivation in China around 10,000 years ago, the planting of rice has spread throughout the world, reaching the USA in 1609 and Australia in 1892. Today rice is the staple food for nearly half the world's people; about 95 per cent of the crop is grown and consumed in Asia, and in China and Japan the word for rice is the same as that for 'meal'.

Yet, while Westerners may associate its daily consumption with many poor countries, rice is traditionally perceived as a luxury food. In Japan, at one stage, it was even used as money. Persian food expert Margaret Shaida says that in Iran, rice is a special food while wheat bread is the 'filler'. 'Certainly in areas where it grows, such as the Caspian, it is part of the staple diet, but most peasants ate rice maybe only once a year, for example at weddings.

Even the well off would only eat it once a week.' This explains the role of expensive or premium ingredients in many Persian rice dishes. 'Foreigners would be given rice when they came to dinner because it was a special occasion, and this has perhaps given the impression that Iranians eat like that more often than they do,' says Shaida. This is true of many other countries, where crops such as millet, barley and maize form part of the everyday diet of working people, but tend not to be featured on restaurant or feast menus.

Variety in the diets of the rice-staple countries of Asia has proved important over and above financial necessity. Rice is nutritious, however over-consumption of milled rice has led to the development of beriberi, a disease caused by a lack of thiamin in the diet, one of the nutrients milled off during processing. Interestingly, while enriching polished rice with vitamins has been explored, no one seems to have taken seriously the option of consuming semi-pearled rice instead.

After the 1950s, overall standards of wealth tended to increase in the Asian countries, prompting a wider intake of foods and improved nutrition, so beriberi receded as a problem. It is noteworthy however that in the rice-growing areas of India, the general health advice in recent years has been to increase fibre consumption by eating wholewheat bread rather than to consume wholegrain rice.

In countries where rice has been adopted, people tend not to want to revert to their previous staples, which may include 'coarser' grains, tubers and root vegetables. The introduction of rice also tends to be commensurate with population expansion. According to the Cambridge World History of Food, rice can support more people per unit of land than either wheat or maize, however history also shows the folly of relying too heavily on one crop. How fortunate we are to have such a wide range of grains.

Rye is
plump, juicy, sweet, dark, chewy, heavy, fruity

It goes with
salmon, ham, cheese, caraway, nutmeg, molasses, cherries

Rye and triticale

Science fiction often says more about the time in which it is written than the time it is written about. In 1967, triticale was cutting-edge, an exciting new grain formed by crossing rye and wheat. A *Star Trek* scriptwriter extrapolated that in future it would be the only Earth grain that could grow on a barren planet called Sherman's Planet, the subject of territorial dispute between the Federation and Klingons. Ensuring the safe transport of the triticale was a vital mission given – naturally – to Captain James T Kirk and the crew of the *Starship Enterprise*.

That episode, 'The Trouble with Tribbles', is one of the most popular in the television series' history and, maybe come Stardate 4523.3, triticale will indeed be a vital food crop. But despite the constant television reruns – and a growing appreciation that science fiction often becomes science fact – few people in the twenty-first century have heard of triticale, let alone eaten it.

'What is triticale?' said one organic farmer I interviewed, who knows exactly what it is. 'Triticale is a confusing message. It's a hybrid. It's not rye, it's not wheat, and it's associated with animal feed, so people switch off mentally.'

That's not how the American Association of General Chemists saw it in 1974. 'A cereal which combines some of the milling and baking qualities of wheat with some of the nutritional characteristics of rye is, at the worst, an interesting development, and at the best a revolution in fulfilling human needs,' their conference papers reported.

Once people try triticale, they seem to like it, especially in porridges, pancakes and waffles. In one study, a Denver-based restaurant chain offered customers pancakes made from wholegrain stoneground triticale flour. In less than a month they took 40 per cent of the chain's pancake business, only to be taken off the menu because of problems securing a regular supply of the flour.

Dr Kath Cooper breeds triticale at Adelaide University, South Australia, and has taken personal responsibility for promoting the grain to farmers and consumers, including writing a cookbook about it. Triticale has become more than a job to her – at home it's also the grain she eats most often. 'I wouldn't keep eating it just because I'm working on it,' she says. 'You have to really like something in order to keep eating it day after day.'

She often has to explain that breeding triticale is not genetic engineering or a clever laboratory technique, just cross-pollination. 'It's a very simple thing to do. I take the pollen off the rye plant and put it on a wheat plant. The resulting plant is naturally fertile and the process could have happened in nature. More intricate breeding goes into wheat these days than has ever gone into triticale.'

Triticale is a naturally sweet grain that does not absorb much water. Cakes and scones made from triticale flour are tender but can dry out quickly. For those people who prefer rye breads made with a portion of wheat flour, triticale conveniently obliterates the need to buy two varieties. The rolled flakes are a pleasantly flavoured addition to oatmeal porridge; used alone their texture might be considered too chewy. In general, triticale can be used anywhere you would use wheat or rye, giving wheat foods a more distinctive flavour and rye foods a milder one.

Intentional or not, an inherent implication of triticale's development is that rye needs improving. Agreed: the whole grains take a long time to cook. However, their sweet juiciness is surprisingly pleasant, the texture hearty yet easy-eating, and they make ideal nourishment for wintry weather. Still, baked goods made from rye flour are the usual way to encounter this underrated grain.

Rye bread in myriad forms is the traditional favourite of northern and eastern Europe – and peasant food if you like that description – but it is an acquired taste associated with gourmet dining in most Anglo-Saxon countries. Germany's dark, dense pumpernickel, which blends rye crumbs or flour with sweet ingredients such as molasses, cinnamon and hazelnuts, is perhaps the most renowned and was originally a way of minimizing wastage by using the coarsest grains. It has many relatives, often flavoured with molasses, but also cocoa, coffee, caramel and caraway. Rye is also made into light, aromatic breads and is particularly good if sourdough starter is used.

In her brilliant book, *The Food and Cooking of Eastern Europe*, Lesley Chamberlain explains why some people may struggle at first to get to grips with the distinctive flavour and texture of rye bread. 'A genuine rye loaf is not a "background" bread like a French stick, but one designed for the prominence of the hors-d'oeuvre or evening cold table, or for eating on its own, East European style, perhaps with yogurt or a glass of buttermilk.'

There is also, she says, an important tradition of using leftover or stale rye bread in other dishes, sweet and savoury. It is grated then layered with chocolate, quark cheese and fruit in Germany, used to thicken spicy sauces in Hungary, made into dumplings or savoury puddings in Bohemia. A great idea from a Southern Australian wine-country restaurant is to soak rye bread in cabernet sauvignon then toast it on a chargrill to serve with pâté.

Cakes made from rye flour are traditional in the Netherlands, and best known today is *ontbijtkoek* or 'breakfast cake'. It is served as a topping on open sandwiches and tastes good with cheese.

Rye flour is low in gluten so do not expect baked goods made from it to rise substantially. Doughs will also be stickier than those made from wheat flours. This is not unappealing, just different. Whole cooked rye berries or flakes are a good addition to grainy breads. For porridges and muesli they are best used with a light hand, in combination with other grains.

Quinoa is
pearlescent, beady, crunchy, bitter, astringent, grassy, mild

It goes with
shellfish, beef, spring onions, soy sauce, vinaigrette, grapes, citrus

Quinoa

Much about quinoa suggests outer space, from the grain's alien shape – which is reminiscent of a tiny UFO or pinhead planet Saturn – to NASA's assertion that it is a super-food worthy of inclusion on long-term missions. Yet while quinoa is being pitched as the nourishment of the future, it is also one of our oldest grains, cultivated by the South American Incas at least 5–8,000 years ago.

The fact that it has been widely unknown in the years hence is due primarily to the Spanish conquistadors, who on their arrival in Peru actively phased out quinoa (and the Incas) in favour of maize, barley and potatoes. It was never a fitting end for the crop known by the locals as the Mother Grain. Legend says that each year the Inca king would plant the first seed of the season using a golden spade. As well as being responsible for the health of the people, he was accountable for the success of the quinoa crop and studied the stars to divine the most appropriate time for planting.

Although the Incas were unable to perform scientific nutritional analyses at the time, their respect for the grain was caused no doubt by quinoa's extraordinary health-giving properties, and its ability to sustain them over generations when, frankly, very little else could be grown high in the Andes. One can only marvel that the Spanish conquistadors chose to promote corn and potatoes in the Old World, but not power-packed quinoa. Such is the depth of distrust they managed to instill in the native population that even today some South American Aymara and Quechua people believe that feeding quinoa to their children will make them stupid, when the truth may be quite the opposite.

Quinoa has a higher protein content than other grains and a superior balance of amino acids that is on par with that of milk and soya beans. It also boasts high levels of calcium, iron, fibre, B-vitamins, vitamin E and phosphorus. However, comparisons

with barley, corn, oats, rice and wheat are not entirely fair, because *Chenopodium quinoa* is in fact a herb seed rather than a cereal grain, and thus related not to grasses but to the broad-leaved goosefoot family, which includes spinach and chard. Being a seed, it's also relatively high in fat and a good source of oil.

Apart from corn, no other grain comes in such a fine range of trendy decorator colours to tempt professional cooks and dinner party hosts. Most supermarkets and health food shops only offer the ivory white or pale gold quinoa but good specialist suppliers will have black and red too. The species also boasts green, pink, orange and purple varieties. Sprout the seeds, if you like, for use in salads, letting them turn green before eating. Distribution of quinoa flour has steadily improved as it has become recognized as a good non-wheat flour. It has a wonderful texture and produces particularly tender pancakes; however you might find that the taste is one that needs to be acquired.

A taste sensation

Some people have described quinoa as the caviar of grains, others less temptingly say it is like a cross between sesame and millet. Certainly once the tiny gelatinous spheres are cooked, the textural association with caviar may seem reasonable, but the slightly acrid and rather hay-like flavour is nothing like beluga, or indeed sesame seed. Caviar also lacks quinoa's crunchy tails, formed when the seed's thin equatorial band (the ring of Saturn) uncurls during cooking. The combination of soft, pearly bead and crunchy spiral tail is at first quite startling, but adds welcome interest to salads, porridges, soups and puddings that most other grains can't match. Also attractive is the fact that, unlike other similarly virtuous wholegrains, quinoa can be cooked in just ten minutes – faster than refined white rice, yet with a similarly light effect on the palate.

Several of the world's most esteemed chefs have taken quinoa to their hearts as well as their kitchens. Chicago-based Charlie Trotter is drawn not only to the grain's nutritional benefits, but to its cooking qualities as well. 'Quinoa is somewhat of a neutral grain, one that beautifully picks up the flavours of the ingredients it is cooked with,' he says. 'It's a great way to introduce flavours to a dish in a not-so-obvious manner. The grain also has a very refined texture which, again, can help add depth or complexity to a dish.'

Another great exponent is Cuban-born Douglas Rodriguez, whose Nuevo Latino cooking style has generated much interest in the traditional dishes and flavours of South America. He makes quinoa soup that is flavoured with shredded chicken, jalapeno chillies, spinach, fresh coriander and mint, as well as a herby quinoa salad teamed with spice-rubbed roast lamb, and grilled octopus with black olive sauce and quinoa.

At New York's Asia de Cuba, which fuses Asian and Cuban cooking, quinoa appears in a salad with tropical fruit, honey, sherry vinegar, parsley and garlic, all topped with a skewer of tamari-marinated beef. Meanwhile, in Europe, chef's chef Ferran Adria of El Bulli, Spain's most prestigious and innovative Michelin-rated restaurant, is incorporating quinoa into his anything-but-traditional degustation menus. Leader of the experimental sci-food trend, he presents a fun and unusual cone of popped quinoa grains to diners that, when eaten, fizzle on the tongue like Poprocks.

It is essential to wash quinoa before boiling it. The seed has a very bitter natural coating called saponin, a resin-like substance that turns water soapy and makes the grain inedible. NASA has speculated that in outer space the saponin washed from the astronauts' hydroponically grown quinoa grains could be recycled into shampoo and detergents. For those of us on Planet Earth the coating is removed during processing, however even clean-looking, well-packaged quinoa will still retain a light coating of saponin dust. Best therefore to put it in a very fine sieve and rinse under running water before use. Bear in mind, however, that the quinoa will always retain a little of its inherent astringency, although many will find this slight bitterness quite pleasant.

Organic cultivation and fair trade

The 500-year decline in quinoa production has reversed over the past decade as South Americans, particularly Peruvians and Bolivians, have developed renewed respect for their indigenous crops and traditional dishes. They have competition, however, for North America has clocked the grain's nutritional value and it's now being grown on the Canadian prairies; as well as in Oregon, the Rocky Mountains, California, New Mexico and Washington in the USA. It's also farmed in Egypt. Impressively, all production is organic. Nevertheless, it is an uncomfortable, but not uncommon, irony that this former staple crop of some of the world's poorest peoples is being grown elsewhere and sold at a premium in gourmet stores for the wealthy palate.

Focusing on this growing and increasingly competitive export market may be an advantageous opportunity to the peasant farmers of the Andes; however another, opposing, view is that they would be better off to keep and consume more of the crop themselves. Quinoa offers far greater nutritional advantages than the imported, subsidized wheat and rice that have become the staples of Peru and Bolivia, and it can be grown locally. Harvesting, threshing and washing is not as mechanized as it is on the large-scale farms of North America, nor the yield as high because of the high wastage of the traditional Andean production methods, thus putting the South Americans at a price disadvantage. Some

Western consumers will happily pay more for Fairtrade-style products that support disadvantaged native communities, however the health food and supermarket trades have a history of severe price consciousness. Fortunately work is being done to help the Andean farmers harvest and clean the grain more efficiently.

Quinoa is relatively easy to grow; indeed, it thrives in mountainous regions, poor soil conditions and very dry climates. You may be able to grow it in your own backyard by paying it very little attention and, if you do, pick the plant's young green leaves to eat raw in salads or cook as for spinach. The ancient South Americans devised sophisticated terracing and irrigation systems to make the most of their rocky environment and low rainfall patterns. In particularly steep areas the strip of terrace available for planting might be as narrow as 15cm/6in. Yet it has been calculated that in ancient times the land was actually five times more productive than it is today, begging a reappraisal of their old agricultural techniques and policies as well as of their indigenous foodstuffs.

As quinoa becomes more popular we may expect a nutritious upturn in our diets, but are faced with one perplexing dilemma: how to ask for it. The correct pronunciation is 'keen-wah', yet say this to most people – including the grain's British importers – and they will look at you strangely. Many English speakers understandably pronounce it as they read it: 'kwin-o-ah', and then once in the habit of that pronunciation find it difficult to change. A solution could be to switch to one of the other South American spellings – quinua – which some people believe is more correct anyway.

Why worry? Because people tend to shy away from foods they are unsure how to pronounce, and quinoa has much to offer, whether as an exotic ingredient for fine dining or as healthy, quick-cooking dietary staple.

A saucepan is all you need to begin cooking with grains. That, plus an active interest in your own eating pleasure and nourishment. Here follows an anthology of some of the world's best grain dishes, with ideas and advice from great cooks all over. Whether you want indulgence or simplicity, fast food or leisurely cooking, purity or pure comfort, there is a dish to suit any mood, any occasion, any lifestyle.

the recipes

Cooking with grains

Store cupboard

Based on sound medical and nutritional advice not to routinely eat the same foods every day, a starting point might be to keep six varieties of grain foods, over and above wheat and rice products, in the kitchen at any one time. That way, even if you cooked all meals at home for two or three days, you need not repeat a grain. Incorporate other starchy foods such as potatoes, pulses and legumes into mealtimes as well and the result is an even more varied diet. A convenient way to ensure variety is to purchase rye, pumpernickel or barley bread for sandwiches and snacks.

Which ones to choose? A mixture of quick-cooking grains and whole grain varieties is important, for practical as well as nutritional reasons. Although a trip to a good organic store may be inspiring, a range of different flours and meals is sensible only if you particularly enjoy baking. Nor is there any point in keeping a bank of speciality wheat products as you are likely to encounter plenty of wheat through bread, pasta, pizza, cakes and so on.

Let's assume that you are au fait with rice and already keep one short- or medium-grain type for making risottos, paellas and so on, plus a long-grain variety to serve plainly boiled or made into pilaffs. Over and above that, my top grain recommendations are stoneground cornmeal, medium oatmeal, buckwheat flour or ready-made noodles, semi-pearled or pot barley, wild rice, and either millet or quinoa.

The ones I most wish I had not left out of this necessarily brief list are semi-pearled farro, freekeh, and buckwheat groats. If I lived in the USA I would be likely to use red posole and black barley more often. As it is they tend to be meted out to honoured guests because they have not to date been easy to buy in London or Sydney. For baking I keep white plain all-purpose wheat flour, plus a wholemeal variety which tends to be spelt rather than common wheat.

Storing

While a key attraction of grains to ancient peoples was their long keeping qualities, modern kitchens tend to run best on a 'little and often' approach to shopping, keeping food storage times to a minimum. This is especially important for stoneground flours that contain the grain's germ and become rancid if kept too long, and items such as couscous, which is truly hideous when stale. Buying grain and grain products in bulk is only worthwhile if you eat a lot of one variety every day, and that is not ideal for health in any case. It's best to stick with several small bags of an assortment of grains. Decant the bags into glass, ceramic, solid plastic or other sealed containers for safe storage and keep in a cool, dark, dry place, or the refrigerator. Barley in particular needs to be stored carefully as it attracts mice.

Equipment

Rice cookers are geared more to the consumption of polished rice and cooking other grains in them is more inconvenient than simply boiling them in a saucepan is.

Nor should you feel the need to buy a pressure cooker. I tend not to use mine except for the rare occasions I cook really, really long-cooking grains such as rye and posole. Similarly, while grains can be cooked in a microwave oven, the cooking time is no different from boiling them.

What I do think is invaluable is a set of cup measures as they make judging portion sizes quick and easy. The item I have been most pleased to acquire while writing this book is a chinois fin. It is designed for sieving sauces but I use it for straining everything because its conical shape directs the flow of water neatly down the sink or into a jug.

If you cook pancakes or griddle breads frequently or in quantity, a large, heavy indoor barbecue plate is much easier to work on than a frying pan, no matter how large, because the sides of the frying pan make flipping more difficult. I would also now not be without a waffle iron, which is much more fun than a rice cooker.

Soaking

Presoaking of grains is not necessary. Some people believe grains need to be soaked in order to reduce their phytic acid levels, which in turn would make the vitamin and mineral content of the grain easier for the body to absorb. Study results vary, however recent research indicates that even under achievable household soaking conditions, the degree to which phytic acid levels are reduced is minimal.

Others believe soaking kick-starts the germinating process, which in turn would make the grain healthier to eat. However, a lot of foods people think are whole grains are not actually 'whole' and consequently are never going to germinate. Another reason to soak is to shorten the cooking time of the grain. This it will do, a little, however you will soon decide for yourself which is more convenient: boiling for an extra 5–10 minutes, or remembering to soak the grain the day before cooking and then carrying through your plan to cook and eat it.

I never presoak grains, except the extraordinarily long-cooking varieties such as posole and rye berries. There are some traditional polished rice recipes for which the grains need to be soaked in order to achieve the right texture in the finished dish. If you are specifically following a recipe for a traditional dish then you should adhere to the soaking instructions given in the recipe.

Cooking times

Refer to the cooking chart (pages 210–211) for general guidance on cooking times, particularly when cooking grains plainly. These are however only approximations. Cooking times vary according to the degree of pearling or processing, the age of the grain, its shape or specific variety, the method of cooking, and what other ingredients are included in the pot. Many manufacturers overstate the length of cooking required on their packaging and while it is worth noting their suggested cooking times, it is also worth testing them 2–3 minutes sooner than the suggested time for quick-cooking grains and 10–15 minutes earlier for long-cooking grains.

The more whole a grain is, the more difficult it is to overcook it, so long-cooking varieties are very forgiving and 5 or 10 minutes here or there makes little difference to the overall eating quality. This is even true of pearled barley. Polished rices, however, need careful attention.

John's granola

While working at Grosvenor House, my friend chef John Campbell won an award from an American communications company for offering the best hotel breakfast in the world, and this granola was part of his menu. Granola began life in the USA as 'granula', an invention of James Caleb Jackson, who baked a wholewheat flour and water mixture in thin sheets, then ground and rebaked them to serve with milk. Dr John Harvey Kellogg took inspiration from this and, around 1878, made a similar product he called granola from wheat, oats and corn.

ingredients

3 tbsp hazelnuts

325g/4 cups rolled oats

125g/1½ cup barley flakes

100g/4 cups malted wheat flakes

30g/⅔ cup wheatgerm

6 tbsp sultanas

3½ tbsp raisins

5 dried apricots, chopped

25g/⅓ cup dried banana chips

3½ tbsp sunflower seeds

350–500g/1–1½ cups runny Mexican honey

serves 15

method

Heat the oven to 180°C/350°F/Gas 4. Place the nuts on a large roasting tray and bake for 20 minutes, or until golden, shaking them occasionally. Meanwhile, combine the grains, fruit and seeds in a large bowl.

Remove the nuts from the oven but do not turn it off. Leave the nuts to cool, then rub off the skins, chop the nuts roughly and add them to the dry ingredients, mixing well.

Spread the muesli out on the roasting tray and drizzle the honey evenly over it. Bake for 20–25 minutes, stirring every 5 minutes to incorporate the honey and break up any large clusters. Remove the tray from the oven and keep stirring the granola every 5–10 minutes until it is cold. Store in an airtight container for up to 2 weeks.

cook's notes

● You want small clusters of granola, not a sticky mass – if the mixture looks more like the latter, simply return it to the oven for further toasting and stirring until the honey is absorbed.

● John is adamant about the use of organic Mexican honey and finds that when using other varieties the recipe does not work as much to his liking. Viscosity is definitely an issue – the thicker the honey is, the more you will seem to require because it won't spread well. Try warming set or crystallized honey before use so that it coats the cereal lightly and evenly.

● Substitute all or part of the honey with maple syrup if desired. Some people prefer a half-and-half combination of honey and an oil such as sesame, corn or sunflower oil, which will lend a fattier flavour to the granola. You could also skip the honey and toasting process altogether, giving a home-made version of packaged muesli.

● When it comes to the dry ingredients, granola is highly flexible. A little more or less of this or that here and there really does not matter, and you should feel confident experimenting. For showing off, think dried cherries and almonds; in tropical mode, choose shredded coconut and dried mango. Chewy medjool dates will lend a healthy toffee-flavoured richness, and sesame seeds a delicious nutty taste. If zen purity is your aim, consider adding roasted and ground soybeans and a few tablespoons of linseeds. Add any fresh fruit, such as raspberries, when serving.

● Replacing the wheatbran with psyllium husks (which are, let's face it, not hugely delicious or easy to eat) is a good way to vary the diet while also making sure that you consume plenty of colon-friendly insoluble fibre.

Oatmeal porridge

Good 'oatmeal' requires meal, not rolled flakes. Whenever someone says 'Wow! That's the best porridge I've ever eaten!', as though delicious porridge were an amazing rather than an everyday thing, it tends to be because the cook or chef has used proper oatmeal. According to Scots food expert Catherine Brown, this means medium oatmeal, sometimes with pinhead (or coarse, steel-cut oatmeal) added to give a rougher texture. 'Fine oatmeal produces too smooth a porridge for Scottish tastes,' she says. Instant oatflakes are just awful.

ingredients

50g/$\frac{1}{4}$ cup medium oatmeal, or a mixture
  of medium and coarse oatmeal
about 300ml/1$\frac{1}{2}$ cups water
a pinch of salt
per person

method

Place the oatmeal a small, heavy saucepan and toast it over a medium-low heat for 2–3 minutes, stirring constantly with a wooden spoon or Scottish spurtle, until lightly fragrant.

Slowly mix in the water (it may spit a little) and salt. Lower the heat right down and cook, stirring occasionally for 20–30 minutes, depending on how crunchy or thick you like the oatmeal. If it becomes too thick, simply add a little more water.

cook's notes

● You'll notice that this mixture is not made with milk or a blend of half milk and half water. The dairy element comes at serving time.

● The 'right' or traditional way to serve porridge in Scotland is to place the cooked cereal in a wooden bowl to keep it hot, and sit a bowl of cold milk or cream alongside. A carved horn spoon is dipped into the porridge, then into the milk or cream, the idea being to enjoy the combination of hot and cold sensations until the last gloop of porridge is eaten.

● Tried and trusted sweeteners for porridge include golden syrup, molasses, maple syrup, honey, brown or white sugar. My friend Keith from Edinburgh likes it with yogurt.

● A bowl of porridge topped with a tablespoon of whisky and a tablespoon of cream is a wonderful thing, especially on winter mornings, but I confess I prefer to use cognac.

● The Scots tradition of leaving excess porridge to set, then cutting it into slices for eating cold or frying as an accompaniment to eggs, bacon or fish is not peculiar to Scotland. These and other ways of using leftover oatmeal can also be found in old American cookbooks. The best is perhaps oatmeal muffins (see cook's notes, page 84).

● At Sarabeth's Restaurants, New York's definitive breakfast venues, oatmeal is a key feature of the menu. Cooked wheatberries are stirred into the porridge to give a crunchy mixture called Big Bad Wolf, which is served with cream, butter and brown sugar. Papa Bear on the other hand has bananas, raisins, cream and honey.

Sticky black rice with coconut and tofu

Tell people you like tofu and they're likely to think you're weird. But they won't have had it freshly made, scooped warm and junket-like from the vat, the soymilk from which it's produced adding just a faint flavour and aroma. This is how I have enjoyed tofu in Japanese Kansai-style restaurants, where it was made in the centre of the table, and at Malaysian market stalls. The following recipe is from the latter, where it was sold with vividly coloured rice porridge and coconut cream as a sweet snack for any time of the day.

ingredients

50g/¹⁄₄ cup black sticky (glutinous) rice

about 300ml/1¹⁄₂ cups water

1 tbsp palm sugar

3–4 tbsp coconut cream, or more as desired

a pinch of salt

100g/3¹⁄₂oz silken tofu

per person

method

Place the rice, water and palm sugar in a small, heavy saucepan and stir to dissolve the sugar. Bring to the boil over a medium-high heat, then reduce the heat right down and simmer for 25–35 minutes, stirring occasionally. Cook for longer if preferred, adding more water to achieve the desired consistency.

Meanwhile, in another small saucepan, beat together the coconut cream and salt until smooth. Place over a low heat and cook gently, stirring occasionally, until warmed through but not boiling.

Place the tofu on a small plate and lay the plate inside a steamer. Set the steamer over a pan of water, bring to the boil, then lower the heat and steam the tofu for 3–5 minutes until hot.

To serve, place the hot tofu in a large bowl. Ladle the black rice porridge over the top and drizzle with the warmed coconut cream.

cook's notes

● The closest approximation your grocer will have to the style of freshly made tofu required here is chilled fresh silken tofu. I can't stand the packaged taste of long-life tofu, no matter how many Oriental people tell me it's good stuff. Never use firm pressed chewy tofu for this dish.

● This recipe is also suitable for black barley, which is available from some specialist grain suppliers and fondly remembered by Southeast Asian expatriates that I've met in London.

● The Malaysian woman at the market stall routinely soaked her black sticky rice overnight before cooking it until almost disintegrated. I tend to prefer the grains still reasonably whole, and have skipped the soaking with no horrifying extension of cooking time. Whatever degree of softness you choose, the mixture retains a slight crunch because of the bran on the rice.

● Sticky black rice and coconut can be combined differently to make a richer porridge suitable for indulgent mornings or desserts. You simply boil the rice in water until tender, then drain and place it in a saucepan with 225ml/1 cup of coconut milk, 1 tbsp palm sugar, 2 tsp white sugar and a pinch of salt. Cook gently until thick, then serve drizzled with a little extra coconut milk or coconut cream.

Egyptian porridge

This is not an authentic Egyptian recipe but based on one in Rena Salaman's inspiring *Healthy Mediterranean Cooking*. It is slightly reminiscent of a Jewish dish called belila, and a Jain one called chunna, which is a speciality of Rajasthan and traditionally served garnished with gold or silver leaf. Even without the garnish, this is an indulgent, sensuous combination, rich enough to be served as a dessert yet deceptively easy to make. The pinky-red pearls of pomegranate make it look stunning and add a welcome note of juiciness.

ingredients

50g/ ¼ cup semi-pearled wheatberries such as farro

40g/ ¼ cup pine kernels

175ml/ ¾ cup milk

2 tsp honey

1 tsp sugar

7 dried apricots, chopped

the seeds of ¼–½ pomegranate

15g/ ¼ cup flaked almonds

1 tsp rose water, or to taste

serves 2–3

method

Place the grains in a large pan of water and bring to the boil. Reduce the heat and simmer for 25–35 minutes, or until tender. Drain thoroughly.

Meanwhile, if desired, toast the pine nuts in a dry frying pan for 3–5 minutes, stirring constantly until golden and fragrant. Remove to a bowl to cool.

Clean out the pan that you used for cooking the wheat. Combine the wheat, milk, honey and sugar in the pan and bring to the boil. Reduce the heat and simmer for 10–15 minutes.

Remove the pan from the heat and stir in the pine nuts, apricots, pomegranate and almonds. Sweeten to taste with rose water. Serve immediately, with additional milk if desired.

cook's notes

● Use wholegrain wheatberries of any type if preferred, boiling them in the pan of water for the necessary extra time to reach tenderness. Large-grained Kamut, derived from an ancient Egyptian wheat, would be an amusing choice, though you might want to chop the grains roughly before combining with the other ingredients as they make the mixture a lot coarser.

● A reasonable version of the authentic Jain dish would be made from wheat, milk, sugar, a little ground cardamom and a few strands of saffron. Using Ebly, Pasta Wheat or any other highly pearled wheatberries rather than wholegrains would be most authentic here.

● Belila, according to Claudia Roden's *Book of Jewish Food*, is made with young wholegrain wheat or pearl barley, and sugar syrup instead of milk. Flavourings would include orange flower water, cinnamon, pistachios, walnuts and soaked raisins or sultanas – you could confidently use any of these in this basic recipe.

● These ingredients can be made into a terrific pudding. Cook the porridge as given here using 200g/1 cup wheat, 180g/1 cup chopped dried apricots, 150g/1 cup pine kernels, 35g/ ½ cup flaked almonds, 900ml/4 cups milk, 2½ tbsp honey and 1 tbsp sugar. Whisk 2 eggs until light and fluffy then slowly mix in a little hot milky porridge. Return to the saucepan and stir gently to combine. Spoon into a greased soufflé dish, sprinkle with a few flaked almonds and bake at 170°C/325°F/Gas 3 for about 30 minutes. Serve sprinkled with pomegranate seeds.

Finnish oven porridge

The Scandinavian method of cooking porridge in the oven is one of the best ways of producing a cooked wholegrain cereal. I include it here because, while it might seem impractical for commuters to make, the many people who now work from home could find that it fits easily into their lifestyle. Put it in the oven on waking, shower, go out and buy the papers and, before you know it, there's a hot, healthy porridge ready to tuck into. In Finland this porridge is called uunipuuro and it is believed to make children grow strong.

ingredients

100g/ ½ cup semi-pearled or pot barley
175ml/ ¾ cups milk
175ml/ ¾ cups water
¼ tsp salt
a little butter and sugar, to serve
serves 2–3

method

Heat the oven to 180°C/350°F/Gas 4. Combine the barley, milk, water and salt in an ovenproof dish. Bake for 1½–2 hours or until most of the liquid has been absorbed and the grains are full blown and tender. Check the mixture about two-thirds of the way through cooking and add some extra milk or water if it is becoming too dry.

Remove from the oven and give the porridge a stir. Ladle into bowls and mix in butter and sugar to taste. Leftovers can easily be reheated for consumption the next day.

cook's notes

● You can easily substitute semi-pearled wheatberries, brown rice or coarse steel-cut oats for the barley in this recipe.

● In Poland a similar and surprisingly tasty dish is made using buckwheat kernels and cream. It includes a dash of vanilla essence, rum and a strip of lemon peel to flavour the mixture, and is best served sprinkled with icing sugar and fresh cherries. These additions really lighten the earthy, meaty taste of the buckwheat.

Bircher muesli

Raw oats for breakfast was not an idea entirely unique to Max Bircher-Benner, the Swiss physician credited with having invented this dish – the Scots beat him to it with their brose. The best-tasting bircher mueslis, especially those served in hotels, tend to include cream, which is not what most people who eat this regularly want in a breakfast. Interestingly, Dr Bircher-Benner's intention was not that his muesli be a high-grain dish, but one that emphasized raw foods, especially fruit and nuts, which he believed made sick people better.

ingredients

2–3 tbsp rolled oats

1–2 tbsp sultanas or raisins

a little water or fruit juice, for soaking

1 apple

a squeeze of lemon juice

2–3 tbsp plain yogurt

1 tbsp chopped unsalted nuts such as almonds,
   brazils or hazels, or a mixture

1 tsp runny honey

a handful of berries, to serve

per person

method

Place the oats and sultanas or raisins in a bowl and cover with a little water or fruit juice. Leave to soak overnight or for several hours to plump and soften.

Just before serving, coarsely grate the apple, leaving the skin on. Stir the apple and the sweet juices it exudes into the oat mixture.

Mix in the lemon juice, yogurt, nuts and honey, then serve the muesli scattered with the berries.

cook's notes

● Grating an apple can seem effort-prohibitive early in the morning, but it is absolutely essential for flavour and juiciness. You could use a pear instead. A generous hand with the berries is another way to ensure a fine result.

● For the creamiest taste choose Greek yogurt or a mild bio yogurt, low-fat if you like. Use acidic tasting yogurt with caution, unless, of course, you prefer it.

● Some people like to add ground cinnamon, ginger or nutmeg; I'm not one of them.

● I often forget to soak the oats overnight and make do by covering them with a little boiling water the next morning. By the time you've grated the apple and mixed in the other ingredients, the muesli is cool enough to eat and the oats are well softened.

● The following idea is from chef Adam Palmer of Champenay's health spa in England, who makes a similar type of muesli with bulgur wheat. He boils 50g/¼ cup bulgur with 5 tbsp sugar, 1 tsp cinnamon, ½ tsp ginger, the zest and juice of an orange, plus 200ml/⅞ cup soya milk, until the liquid has been absorbed and the grains are tender. Then he stirs in almonds, diced red and green apples and pears, and tops this mixture with a simple blackcurrant and honey compote. Very glamorous.

● To make Scots brose, place a portion of oat, barley or pease meal in a heatproof bowl and cover with hot milk or water. Stir in some salt and a little butter, plus dried fruit if desired. It is also traditional to use leftover cooking water from meat or vegetables in this dish.

Griddle cakes

Terrific pancakes drenched in maple syrup, and preferably served with bacon, are for many of us the culinary highlight of trips to the USA. We have the Dutch to thank for taking pancakes, as well as waffles and doughnuts, to the New World, but we also have them to blame for coleslaw. Much as I love pancakes, these days I find them flabby and bland when made from only white wheat flour; a proportion of buckwheat or cornmeal is almost compulsory for me, unless the recipe is particularly rich in eggs or features ricotta or curd cheese.

ingredients

75g/ $1/2$ cup buckwheat flour

50g/ $1/3$ cup plain wheat-type flour

2 tsp baking powder

1 tsp sugar

$1/4$ tsp salt

3 eggs, separated

150ml/ $5/8$ cup milk

1 tbsp butter

a little butter or vegetable oil spray, for greasing

serves 3–4

method

In a large bowl, combine the buckwheat flour, wheat flour, baking powder, sugar and salt. Stir to combine, then make a well in the centre.

Drop in the egg yolks and begin to slowly mix in the flour mixture. Gradually add the milk, stirring to a smooth batter, then set aside for 20–30 minutes.

Meanwhile, melt the butter ready to add to the batter. When almost ready to cook, whisk the egg whites in a large bowl until stiff peaks form.

Fold a large spoonful of the egg whites into the batter to lighten the mixture, then fold in the remaining egg white and melted butter.

Heat a griddle or large heavy frying pan over a medium-high heat. Use a small ladle to add equal quantities of batter to the pan to make round pancakes. When bubbles have appeared on the surface and then popped, and the underside is a rich golden brown, flip the pancakes over and cook on the other side for 2–3 minutes. Serve immediately, or keep warm on a tray in a low oven while you finish cooking the rest of the batter.

cook's notes

● This isn't so much a recipe as a formula. You can use pretty well any grain flour or mixture of flours that you like in it, provided you stick with these basic proportions of wheat flour and complementary grain. You can also add 15g/ $1/2$oz of a cooked grain such as millet or wild rice to the batter with no ill effects.

● To make blueberry or raspberry pancakes, simply press the fruit into the soft top side of the pancakes just after they go on the griddle.

● I have found this recipe, when made with buckwheat, vastly superior to any authentic yeasted blini recipe I've tried, and much easier to cook. Don't hesitate to use this one when serving canapés of tiny pancakes topped with soured cream and smoked salmon or fish roe.

● For days when you don't fancy separating eggs and whisking the whites, simply stir one whole egg into this batter; it will then feed only two, but it makes pancake day a real no-brainer.

Breton crêpes

Crêpes are often considered to be difficult to make, yet they are simple and demand merely that you observe a few rules. A non-stick pan really helps, so does a good bit of wrist action while pouring the batter, which must rest for 30 minutes before cooking. You might want to use clarified butter for greasing the pan – always, always lightly – as it will prevent scorching. Whatever you do, the first one is unlikely to work out. Don't question why. It's like the changing of the tides or the waxing and waning of the moon. Just accept it and let it be.

ingredients

50g/⅓ cup plain wheat-type flour

50g/⅓ cup buckwheat flour

a pinch of salt

2 eggs

225ml/1 cup milk

2 tbsp butter

a little butter or vegetable oil spray, for greasing

serves 4

cook's notes

● Lemon juice and sugar are for me the best accompaniments by far. For a hearty savoury dish, break an egg onto the uncooked side while the crêpe is cooking, and let it 'fry'. When just set, sprinkle with a little finely grated gruyère cheese and some pepper, and roll up.

● Decreasing the milk in this recipe by 3 tbsp will give a thicker pouring batter suitable for batter puddings such as clafoutis, Yorkshire pudding and toad-in-the-hole.

● Buckwheat-based batter is especially good for clafoutis, which is traditionally made by laying a cup or so of griottines (marinated sour cherries) in the base of an ovenproof dish and covering them with the batter. A terrific alternative is 5 quartered plums marinated with 4 tbsp of maple syrup. Bake at 180°C/350°F/Gas 4 for about an hour.

● For very delicate lacy crêpes, use cornflour (cornstarch) in place of the buckwheat flour.

method

Sift the wheat flour, buckwheat flour and salt into a large mixing bowl. Make a well in the centre and break in the eggs. Whisk until the eggs are broken up thoroughly. Gradually add the milk, whisking to a smooth batter, then set aside for 30 minutes.

Meanwhile, melt the butter and stir it into the batter just before cooking.

Place a steel or non-stick crêpe pan (about 18cm/7in in diameter) over a medium-high heat until it is very hot. The pan is ready for cooking when a drop of water bounces on the surface. If the water just lies there, the pan is not hot enough; if it quickly evaporates it is too hot.

When ready, lightly grease the pan. Lift the pan off the heat with one hand and use the other to pour a small ladle of batter into the pan, while turning and tilting the wrist of the other hand to ensure the batter spreads lightly and evenly over the base of the pan. Really, it's not that hard.

Cook the crêpe for 1–2 minutes or until the top side is set and the underside is browned. Using a palette knife, lift up the side of the crêpe and ease the palette knife under. Flip over the crêpe and continue cooking on the other side until done. Set aside on a plate while you cook the rest of the batter, stacking the crêpes as you go.

Wild rice waffles

These are the best waffles I've ever eaten, bar none. The idea came from one of my all-time favourite cookbooks, *Good Mornings*, by American food writer Michael McLaughlin. In it he meticulously plans what steps to take the night before to ensure all his wonderful dishes are quickly and easily prepared in the morning, suggesting that he must be a lovely man to wake up with. Indeed, I can confirm the pulling power of these waffles and have become so enthused of my waffle machine that I'd love to trade up to a higher-spec, more expensive model.

ingredients

75g/ ⅓ cup wild rice

1½ tbsp butter

100g/ ⅔ cup plain wheat-type flour

50g/ ⅓ cup coarse yellow cornmeal

1 tbsp sugar

1 tsp baking powder

a large pinch of salt

1 egg

175ml/ ¾ cup milk

a little vegetable oil spray, for greasing

butter and maple syrup, to serve

makes 4

cook's notes

● Michael McLaughlin's recipe uses buckwheat flour instead of cornmeal, buttermilk in place of milk, and oil in place of melted butter. Buttermilk is not easy to get in my area, so I use regular milk. You could sour it with a teaspoon of lemon juice if desired.

● These waffles are good with black barley, millet or uncooked amaranth mixed in instead of the wild rice, but for me the wild rice version is definitely superior.

● Most pouring batter recipes, sweet or savoury, are suitable for cooking as waffles. The key difference between waffles and thick pancakes is that waffles require a little more fat to make them crisp.

● I've tried yeasted waffle batters and can't see the point of them. The batter doesn't need to puff up only to be squashed down again by the iron. All it does is make the waffles taste yeasty.

method

Place the wild rice in a small saucepan, cover with a generous quantity of water and bring to the boil. Reduce the heat and simmer for 45 minutes, or until the grains are butterflied and tender. Drain and rinse under cold running water. Set aside to drain thoroughly.

Melt the butter and set aside. Sift the wheat flour, cornmeal, sugar, baking powder and salt into a large mixing bowl. Make a well in the centre and crack in the egg, whisking to break it up. Gradually whisk in the milk and melted butter to give a smooth batter. Stir in the cooked rice.

Heat a waffle iron, preferably non-stick and electric, according to the manufacturer's instructions. Coat lightly with vegetable oil spray. Pour a large ladle of batter into the machine, or enough so that it is about three-quarters full, and clamp shut.

Cook for 3–4 minutes or until the waffle has browned and crisped to your liking. Keep warm in a low oven while you finish cooking the rest of the batter. Serve hot, topped with butter and maple syrup.

Dutch honey cake

'Cake' here will be a misnomer to anyone familiar with the process of creaming butter and sugar until light and fluffy, then beating in an egg, plain flour and milk. This extraordinarily quick mixture has no butter, no eggs and uses rye flour. The effect is reminiscent of malt loaf rather than madeira cake, so the recipe features in this section accordingly. In the Netherlands it is traditionally eaten for breakfast with cheese and bread, something that will appeal to anyone who enjoys the combination of fruit cake and Wensleydale.

ingredients

300g/2 cups rye flour

200g/1 cup dark brown sugar

1 tsp baking powder

1 tsp ground cinnamon

½ tsp grated or ground nutmeg

½ tsp ground cloves

225ml/1 cup milk

serves 10

method

Heat the oven to 180°C/350°F/Gas 4 and grease a loaf tin. Sift the rye flour, brown sugar, baking powder, cinnamon, nutmeg and cloves into a large mixing bowl and make a well in the centre.

Gradually stir the milk into the dry ingredients to give a smooth, thick batter. Pour the batter into the prepared loaf tin and bake for 60–75 minutes, until a skewer inserted in the centre comes out clean.

Remove from the oven and leave to stand in the tin for 15 minutes before turning out onto a wire rack to cool further. Wrap the cooled loaf in foil to store.

cook's notes

● Unlike other quick breads, but in common with many rye products, this loaf tastes better a few days after baking. Indeed, the extent to which it improves is quite astonishing.

● For a ginger version of this loaf, add a little powdered ginger, plus 4–6 tbsp chopped glacé, crystallized or drained preserved stem ginger just before it goes into the loaf tin.

● For a fruity version, stir in 4 tbsp sultanas and 4 tbsp finely chopped dates.

● This loaf is terrific spread with butter, and a great recipe to know when catering for people with wheat allergies. You could also, if necessary, make it with water in place of milk.

New Orleans pain de riz

'It has been said, and justly, that the only people who know how to make corn bread are the Southern people,' says *The Picayune Creole Cook Book*, a collection of Louisiana-French recipes now over 100 years old. With characteristic pride it states: 'The Creoles, like all true Southerners, never use the yellow cornmeal for making bread, but always the whitest and best meal. In the South the yellow meal is only used to feed chickens and cattle.' This old-time bread is intended for breakfast, but in contemporary terms, would be better at brunch .

ingredients

150g/1 cup white cornmeal

1½ tsp baking powder

½ tsp salt

90g/½ cup cooked white rice

2 eggs

340ml/1½ cups milk

½ tbsp butter, melted, plus extra for greasing

serves 10

method

Heat the oven to 200°C/400°F/Gas 6 and grease a 20cm/8in square cake tin. Sift the cornmeal, baking powder and salt into a large mixing bowl. Rub the cooked rice into the meal until it resembles fine breadcrumbs. Make a well in the centre of the bowl.

Beat the eggs together in a jug and mix in the milk and melted butter. Stir the liquid ingredients into the mixing bowl, beating well to give a light batter.

Pour the mixture into the prepared tin and bake for 30 minutes. Remove from the oven and serve hot with lots of butter and preserves, or alongside savoury foods with plenty of sauces and relishes.

cook's notes

● Despite the richness of this mixture, which features eggs, milk and butter, it is like many cornbreads in that it can be disconcertingly dry to those unused to the texture. That is why it is especially important to serve it with suitably saucy accompaniments and lots of butter.

● I've presented this corn bread to guests with sausages, bacon, eggs and mushrooms plus plenty of home made tomato sauce (page 200). Baked beans (see recipe for tomato sauce, page 200) would be good too, as well as slices of roasted red bell pepper.

● Though I have not yet tried it, roast chicken and a creamy gravy seem to me to be ideal partners for this bread too.

● You could divide the batter between muffin trays or pans and bake them for 20–25 minutes to give individual corn breads.

Orkney-style bannocks

To many people living outside Canada it is surprising to learn that the historic Scots bannock is seen as good traditional fare of the First Nations peoples. They have been eating the simple unleavened griddle-baked bread, made with cornmeal rather than barley and wheat flours, since the seventeenth century when Scottish fur traders took the method to North America. Like Australia's damper and Africa's roosterkoek, the bannock requires few ingredients and can be easily cooked outdoors on a hot plate as well as baked in an oven.

ingredients

300g/2 cups bere meal or barley flour, plus
    extra for dusting
150g/1 cup plain wheat-type flour
1 tsp baking soda
1 tsp cream of tartar
about 225ml/1 cup buttermilk
a pinch of salt
makes 4

cook's notes

● If baking, heat the oven to 190°C/375°F/Gas 5 and bake for 20 minutes.

● You can use fine oatmeal or oat flour in place of the bere meal, or make the mixture entirely from fine white cornmeal.

● Some people like to work in 25g/2 tbsp fat such as butter or lard, or a combination of the two. Rub them into the dry ingredients if using. The mixture can also be sweetened with a little treacle if liked.

● Serve with lots of butter and cheddar cheese and a glass of beer.

method

Place a griddle or heavy-based frying pan over a medium to low heat and leave to get hot.

Sift the bere meal, plain flour, baking soda and cream of tartar into a large bowl to aerate the mixture, then stir in any remnants of bran from the bere meal. Make a well in the centre of the dry ingredients and work in just enough buttermilk to give a stiff dough.

Dust a work surface generously with flour and tip the dough out onto it. Divide the mixture into four even pieces and roll out each one in turn to give rounds about 3.5cm/1½in thick. Place on the hot griddle and cook for 2–3 minutes on each side until the crust is browned and the inside is cooked. Serve immediately.

Multigrain scones

Most wholemeal scones are pretty dire, especially to anyone who believes that the scone's rightful home is nestled under billowy spoonfuls of clotted cream and strawberry jam. If you particularly want to add fibre or other nutritional benefits to the dough, there are better ways to go about it than switching to wholemeal flour, and this is a good example of one. Fierce debate rages (still) in Britain over the proper pronunciation: 'skoan' or 'skon'. As it is originally a Scots word, surely we must defer to them. Skon it is.

ingredients

1 tbsp whole or semi-pearled wheat-type berries
  such as farro
2 tbsp wild rice
225g/1½ cups plain wheat-type flour
2 tsp baking powder
½ tsp salt
50g/¼ cup butter, diced
4 tbsp caster or granulated sugar
1 tbsp millet
120ml/½ cup buttermilk
1 small egg
makes 12

cook's notes

● Traditionally scones were cooked on a griddle as for bannocks, and in Scotland were often made using barley or oat meals.
● Pumpkin scones are renowned in Australia. To make them, add 150g/½ cup of well-mashed steamed pumpkin to the dough.
● Currants and sultanas are common additions to the mixture, but arguably better are chopped pieces of good-quality dates. Dried cherries, blueberries and cranberries are also good. Use 60g/⅓ cup fruit for this quantity of dough. Choc chips are also sometimes added but their taste and texture don't work well here.
● For savoury scones, mix in 40g/⅓ cup of strongly flavoured grated cheese, and fresh herbs if you like. Reconstituted, finely chopped dried mushroom plus spring onion is a good combination for scones to serve with soup.

method

If using wholegrain wheatberries, cook them in a small saucepan of boiling water for 1–1½ hours, or according to the packet instructions, until tender. If using semi-pearled wheat, cook it in the same pan of water as the wild rice, for 45 minutes. Rinse and drain thoroughly, then spread out on a plate to dry. Heat the oven to 230°C/450°F/Gas 8.

Meanwhile, sift the flour, baking powder and salt into a large mixing bowl. Rub in the butter with your fingertips until the mixture resembles fine breadcrumbs, then stir in the sugar, raw millet and the cooked wheat and wild rice. You should be adding about 6 tbsp of grains in total.

Mix the buttermilk into the other ingredients to give a soft dough. Dust a work surface with flour and tip the dough onto it. Knead the dough briefly, then roll out the dough into a square 3.5cm/1½in thick. Trim the edges and cut a grid into the rolled dough to give 12 squares.

Dust a baking sheet with flour and nestle the scones close together on it. Beat the egg in a small bowl and brush it lightly over the tops of the scones to glaze. Bake for 20 minutes or until the scones are risen and golden. Serve warm.

Oatmeal quick bread

This really is a 'quick bread', something you can mix up from store cupboard ingredients in a matter of minutes and put straight into the oven to serve hot alongside supper dishes. The arrival of an unexpected, freshly made loaf at the table tends to bring cries of delight and admiration; the result is devoured so greedily that there are few leftovers for people to eat on subsequent days, when they will notice that the texture is less good. This recipe is essentially a large unsweetened scone and, like them, needs to be enjoyed the day of baking.

ingredients

300ml/1¼ cups milk, plus extra to glaze

a squeeze of lemon juice

100g/⅔ cup plain wheat-type flour,
*   plus extra for dusting*

2 tsp baking powder

½ tsp salt

100g/1½ cup medium oatmeal

a little milk, to glaze

serves 6

method

Heat the oven to 200°C/400°F/Gas 6. Combine the milk and lemon juice in a jug and set aside in a warm place for 10 minutes or so.

Sift the plain flour, baking powder and salt into a large bowl. Stir in the oatmeal, then add the soured milk and quickly mix to give a soft dough.

Dust a work surface with flour and tip the dough out onto it. Knead briefly to give a smooth ball, then shape into a round loaf.

Dust a baking sheet with flour and place the dough at the centre. Brush it lightly with milk and bake for 20 minutes or until the bread is risen and golden and sounds hollow when tapped on the base.

Remove the bread from the oven and wrap it in a clean tea towel until ready to serve.

cook's notes

● To make a herb-flavoured bread, mix in 1½ tbsp of chopped fresh herbs, choosing robust varieties such as thyme or rosemary.

● For a bread that breaks off into sections or triangular farls, cut the round of dough with a large knife and place the dough pieces together on the baking tray. Alternatively you can shape the mixture into individual rolls and bake them for 15 minutes.

● Buttermilk can be used in place of the soured milk and might even be considered preferable. You could also sour the milk with a tiny splash of vinegar if necessary.

● Rolled flakes of oat, rye, wheat or barley make good alternatives to the oatmeal. You could also mix in a few spoonfuls of any leftover cooked grains such as wild rice, barley or wheatberries you happen to have in the house, and a little raw millet if desired.

Claire's Mediterranean corn bread

White wine is the brow-raising ingredient in this unusually flavoured corn bread from my friend Claire Clifton, who began making it regularly after reading an article by British food writer Simone Sekers. For those with a taste for things salty and piquant, people who feel their 'downfall' is olives or cheese, this is a treat along the lines of a savoury Mediterranean cake. It makes a great teatime snack or soup and salad accompaniment, and could be served as part of a bread basket for an Italian meal. The leftovers are super when toasted.

ingredients

250g/1⅔ cups fine cornmeal

3 tsp baking powder

4 eggs, beaten

200ml/⅞ cup olive oil, plus extra for greasing

200ml/⅞ cup white wine

150g/1 cup pitted sliced olives

50g/½ cup pine kernels

3 tbsp chopped roasted bell peppers
*   or sun-dried tomatoes*

serves 6–8

method

Heat the oven to 200°C/400°F/Gas 6 and lightly oil a 23cm/9in spring-clip cake tin. In a large mixing bowl, combine the cornmeal and baking powder and make a well in the centre.

Using a whisk, beat in the eggs. Combine the oil and wine in a jug, then whisk the mixture into the batter. Fold in the olives, pine kernels and roasted bell peppers or sun-dried tomatoes until well distributed.

Pour the batter into the prepared tin and bake for 1¼ hours or until a skewer inserted in the centre of the cake comes out clean. Serve warm.

cook's notes

● The assumption is that you are drawing on a larder containing jars of the preserved antipasto items that are so popular today and easy to buy ready-made. You could, if necessary, roast your own peppers (see page 200). If you are using sun-dried tomatoes packed in oil, remember to pat them dry. If they are not packed in oil, reconstitute them in hot water, rinsing well to remove any excess salt.

● You can mix in pretty much whatever flavourings you like – chillies, capers, chopped ham, cheese, anchovies, walnuts – but keep the total weight of added ingredients to 250g/9oz and restrict any pungent items such as capers or anchovies to very small proportions. Three would sensibly be the maximum number of flavourings to add, and 2 (rather than being a compromise) is very good.

● I have only made this with cornmeal but Claire has used a mixture of cornmeal and wheat flour, and has also produced a wholewheat version that she reports was satisfyingly light. Given the Mediterranean flavourings featured, yellow cornmeal would perhaps be the colour of choice, but it doesn't really matter.

Americas bread

This attractive, seed-studded, granary-style loaf combines the best of the American grains and adds another staple from the New World that the Old World cherishes, the potato. Although rich in wholesome natural ingredients, the finished bread is elegant and light, good for serving at stylish informal lunch parties, or when making sandwiches or toast. Despite what may seem like many steps and components, it's actually gratifyingly easy. The recipe is based on one from esteemed baker and caterer Beth Hensperger.

ingredients

2 tbsp wild rice

175g/6oz potato, peeled and cubed

4 tbsp quinoa

1 tbsp butter

2 tbsp cornmeal

125ml/ 1/2 cup milk

375g/2 1/2 cups plain wheat-type flour, plus
   extra for dusting

75g/ 1/2 cup wholemeal wheat-type flour

2 tsp active dried yeast

1 tbsp caster or granulated sugar

1 1/2 tsp salt

1 1/2 tbsp finely chopped pumpkin seeds

2 tbsp popped amaranth (see cook's notes)

4 tbsp rolled oats

vegetable oil, for greasing

makes 1 large loaf

method

In a small pan of boiling water, cook the wild rice for 45 minutes or until tender. Meanwhile, in separate pans of water, boil the potato for 20 minutes or until tender, and simmer the quinoa for 12–15 minutes.

When cooked, drain the grains and spread them out on a plate to dry. Drain the potato, reserving 175ml/¾ cup of the cooking water. Mash the potato with the butter. Place the cornmeal in a small bowl and stir in the reserved hot cooking water.

Heat the milk gently until lukewarm. In a large mixing bowl, combine 150g/1 cup of the plain flour with the wholemeal flour, yeast, sugar, salt and pumpkin seeds. Work in the mashed potato and wet cornmeal, then add the warm milk, popped amaranth, cooked wild rice and quinoa and mix well.

Gradually add the remaining 225g/1½ cups, or more, of plain flour as required to give a soft, but not sticky dough. Dust a work surface with flour and knead the dough for 5 minutes until elastic.

Wash and dry the mixing bowl and grease it with oil. Place the dough in it, then turn it over so that it is covered in oil. Cover with clingfilm and leave to rise for about 1½ hours or until the dough has doubled in volume.

Oil a loaf tin and dust the inside with the rolled oats. Punch down the risen dough, knead well, then shape into a loaf and lay it in the tin. Leave to rise for 40 minutes until just above the rim of the tin.

Heat the oven to 200°C/400°F/Gas 6. Bake the bread for 10 minutes, then reduce the heat to 180°C/350°F/Gas 5 and continue cooking for a further 30–35 minutes until the loaf is browned and sounds hollow when tapped on the base. Cool on a wire rack.

cook's notes

● To pop the amaranth yourself, add 1 tbsp of seeds to a hot frying pan and stir constantly with a natural-bristled brush until they pop, being careful not to let them burn (see cook's notes, page 192).

Soweto grey bread

Grey rather than brown, this partially wholemeal loaf from South Africa is a good stepping stone to full-on wholemeal bread for those reluctant to make the move. It's also a good first-time yeast bread for anyone new to baking. The quantity of yeast is comparatively high so the bread rises well and quickly, making the proving process seem less intimidating to fledgling breadmakers. This short-sharp-shock method is treated with disapproval by some bakers, but is beneficial in the hot climate from which the recipe originates.

ingredients

250g/1⅔ cups plain wheat-type flour,
    plus extra for dusting
100g/⅔ cup whole wheat-type flour
10g/⅓oz active dried yeast
1 tsp caster or granulated sugar
1 tsp coarse sea salt, plus extra to glaze
225ml/1 cup lukewarm water
1½ tbsp butter, plus extra for greasing mixing bowl
2–3 tbsp poppy seeds
1 tsp cornflour (cornstarch)
makes 1 large loaf

cook's notes

● The poppy seeds can be omitted, in which case the dough simply needs to be kneaded again and then shaped as desired. You could also use sesame seeds in place of poppy.

● Alternatively, mix the poppy seeds into a chopped onion fried in butter until soft and golden. Cool and spread this mixture over the uncooked dough before rolling it up.

method

In a large bowl, combine the two flours, yeast, sugar and salt. Mix in the water and work into a dough, adding more water as necessary to give a soft, but not sticky dough.

Grease your palms well with some of the measured butter and knead the dough in the bowl for 5 minutes. Butter your hands again and knead the dough some more. Continue buttering your hands and kneading until the measured butter is used up and you have been kneading for about 15 minutes.

Grease the mixing bowl with additional butter. Shape the dough into a ball, place it in the bowl, then turn so that the top is greased. Cover with clingfilm and leave to prove in a warm place for 10 minutes; the dough should be allowed to rise, but not so much that it doubles in volume.

Punch down the dough, then knead it again briefly. Turn the dough out onto a floured surface and roll it into a square. Sprinkle the poppy seeds over the top of the dough and press them in very gently. Carefully roll up the dough like a Swiss roll to enclose the poppy seeds. Cover with clingfilm and leave in a warm spot until doubled in size, about 20 minutes.

Heat the oven to 200°C/400°F/Gas 6 and place the loaf on a floured baking sheet. In a mixing bowl, combine the cornflour with 1 tsp salt and 2 tsp water, stirring to make a paste. Brush this over the top of the loaf and bake for 5 minutes, then lower the heat to 180°C/350°F/Gas 4 and continue cooking for a further 40 minutes until the loaf is browned and well risen.

Norwegian flatbrød

Flatbrød is the national bread of Norway. It is a thin, crisp, cracker bread made from flours such as rye and barley, and will seem familiar to nearly everyone who has been on a diet. As with the oatcakes of Scotland and Ireland, once flatbrød is cooked it lasts a very long time, even years. This was a traditional necessity in mountainous Norway, where the geography and climate made agriculture and transport difficult for much of the year. Although eaten daily, breads such as this would be made only a few times annually.

ingredients
50g/ 1/3 cup barley flour, plus extra for dusting
50g/ 1/3 cup rye flour
50g/ 1/3 cup fine oatmeal
1/4 tsp salt
125–150ml/ 1/2 –2/3 cup water
makes 6–8

method
Combine the barley flour, rye flour, oatmeal and salt in a large bowl. Make a well in the centre and gradually add the water, using only just enough to make a workable, but sticky dough.

Dust a work surface with flour and briefly knead the dough into a smooth ball. Divide it into 6–8 pieces depending on the size of your griddle or frying pan.

Place a griddle or large, heavy frying pan over a medium to low heat and leave until very hot.

Roll out each piece of dough very thinly into a large circle. Trim the edges to neaten if desired. Place a circle of dough on the griddle and cook very slowly for several minutes until quite crisp on the bottom, then turn over and cook the other side. Press down on the flatbrød with a fish slice if necessary to keep it flat against the hot surface. Reduce the heat if the bread seems to be browning rather than crisping.

Remove from the griddle and repeat with the remaining rounds of dough.

Eat warm or leave to cool and store in an airtight tin.

cook's notes
● Milk can be used in place of water.
● Another typical recipe features wholemeal wheat flour in place of the fine oatmeal. You could of course use spelt or Kamut flour. Other Norwegian flatbrøds include mashed potato.
● The rolled mixture can be cut into pieces and baked on a tray at 180°C/350°F/Gas 4 until crisp if preferred.
● When rolling the dough, remember the thinner, the better, to achieve the right texture.

Corn tortillas

Mexican food is the most popular ethnic cuisine in the USA, but in the rest of the world is terribly underrated. The main reason is that elsewhere we rarely get to experience good Mexican restaurants, there being no substantial immigrant population, and many of the venues selling 'Mexican' food are rowdy joints geared more to the consumption of tequila slammers than authentic regional cooking. Fresh cornmeal tortillas are almost unheard of, yet there are signs of hope, and the delightful masa harina is increasingly easy to purchase.

ingredients

250g/1⅔ cups masa harina, plus extra for dusting
about 225ml/1 cup hot water
makes 12

cook's notes

● A tortilla press (a Spanish invention) is helpful, but not essential. Mexican cookery experts simply work the dough ball between their palms to flatten it. The rest of us can get by with a heavy frying pan and firm intentions.

● Authentic corn tortillas are smaller than the wheat flour tortillas sold for Tex-Mex cooking. You will need to serve 2 per person.

● For a terrific quick meal, fill the tortillas with a mixture of 300g/10oz boiled potatoes, cubed and sautéed in a little oil with 150g/5oz chopped chorizo, some dried oregano and salt.

● Another first-class treatment is to fill them with fresh white crab meat and tomato salsa. Make a sauce by puréeing 150g/1 cup pumpkin seeds with 225ml/1 cup fish stock until very smooth and thick. Heat the sauce gently in a pan then spoon it over the warm filled tortillas.

method

Take a thick plastic food storage bag and cut around the seams with scissors to give two tough sheets of plastic. Set aside until ready to press the tortillas.

Place the masa harina in a large bowl and make a well in the centre. Pour in most of the hot water and stir to incorporate the flour and give a soft, but not sticky dough. Add more water if necessary.

Dust a work surface with extra masa harina. Turn the dough out onto it and knead lightly 5 to 6 times. Divide the dough into 12 equal pieces and shape each one into a ball. Cover them with a damp tea towel.

Take a ball of dough and place it between the sheets of plastic. Lay them in a tortilla press and close it. Move the handle from side to side to help flatten the ball until it is about 13cm/5in across. Open the press and turn the dough halfway round, then close and flatten again. Remove the tortilla from the plastic, retaining the plastic for the next one.

Place a griddle or large, heavy frying pan over a medium heat. When it is solidly hot, lay the tortilla on it and cook for 30 seconds. Flip it over and continue cooking on the other side for another 30 seconds or until little brown spots start to appear underneath. Flip it over again and continue cooking for a further 30–45 seconds, at which point the tortilla should puff up a little. Remove from the heat and keep warm while you press and cook the remaining tortillas.

Millet congee

One of the nicest things about being in the Orient is the opportunity to enjoy traditional breakfast dishes, of which congee is one. Usually it is made from rice, although millet congee is popular in Taiwan, where it is also a street food snack – I'm told that a good version can be found at the guotie place on the corner of Tienmu West and Tienmu North roads. Congee requires time to cook but very little effort – just an occasional check to ensure the pan hasn't boiled dry. It makes a good, healthy late breakfast for those of us who work from home.

ingredients
100g/ 1/2 cup millet
570ml/2 1/2 cups chicken stock
a few chopped pickled chillies
a little chopped spring onion
a few drops of soy sauce
a few drops of sesame oil
a few fresh coriander leaves
serves 2–3

method
Place the millet in a large, heavy saucepan over a moderate heat and toast the grains lightly for 1–2 minutes, stirring constantly.

Pour in the stock and bring to the boil. Half-cover the saucepan, then reduce the heat and cook gently for 1–2 hours, stirring occasionally, until the mixture has thickened and the grains have burst. Add more stock or some hot water from the kettle if the congee seems to be reducing too quickly.

Serve the congee in large bowls, sprinkled with the chopped chillies, spring onions, soy sauce, sesame oil and coriander leaves (plus any of your preferred optional extras) to taste.

cook's notes
● Millet gives congee a pale yellow colour and vaguely crunchy texture. You can use this basic method to make a white rice congee as well. Cooked for this length of time it will have a smoother and even more soothing texture than the millet version and a slightly milder flavour.

● As with other porridge dishes, the correct serving consistency is the one that you prefer.

● Optional extras to stir into the congee shortly before serving include slices of char siu (Chinese barbecue roast pork), which can be bought ready-cooked from Chinese stores, some cooked chicken or prawns, salted peanuts or a little fresh grated ginger. Sprinkle in a few drops of fish sauce if you like it.

● Don't try and use any of the Chinese coloured rices for congee – when cooked to a pulp, they look highly unappetizing.

● Australian food writer Jill Dupleix makes the excellent suggestion of serving congee as a late supper dish with a glass of champagne.

Quinoa, chicken and spinach soup

Chef Douglas Rodriguez is one of my favourite restaurateurs and food writers, primarily because his inspirational 'Nuevo Latino' style combines both common and exotic ingredients using unfamiliar but generally simple techniques. This recipe is based on one in his book *Latin Ladles*, and treats quinoa as it needs to be treated, with the sweetness of the chicken stock, chillies and herbs off-setting the natural bitterness of the grain. It's great 'flu' food and, although the taste and texture change slightly in the freezer, good to make and store in portions.

ingredients

1 medium whole chicken

2 onions

3 stalks celery

2 carrots, roughly chopped

1 bouquet garni

5–7 cloves garlic

a large bunch of coriander

6 sprigs parsley

3 tbsp softened butter

500g/1lb 2oz baby spinach leaves

leaves from 6 sprigs of mint

3 jalapeno chillies, finely chopped

150g/ ¾ cup quinoa, rinsed thoroughly

5 spring onions, finely chopped

salt and pepper

serves 6–8

cook's notes

● White rice or hominy are good alternatives to quinoa in this recipe.

● For a quicker version of this recipe, use ready-made stock and 4–5 chicken thigh fillets. Cook them in the stock for 20–25 minutes with the herb stalks before picking up the recipe at the point where the vegetables are sautéed.

● I like to add some finely sliced red chillies to this soup for their attractive colour contrast.

method

Rinse the chicken and place it in a large stockpot with 2 litres/9 cups water. Halve one onion and roughly chop 2 stalks of celery and add them to the pot, together with the carrots, bouquet garni, 4 unpeeled garlic cloves and the stalks from the coriander and parsley.

Bring to the boil, then half-cover the pot, reduce the heat right down and simmer for 1 hour or until the chicken is done.

Carefully lift out the cooked chicken, allowing the liquid in the cavity to run back into the pot. Set the chicken aside to cool. Strain the stock, discarding the solids, and rinse out the pot.

Finely chop the remaining onion, celery and garlic. Melt the butter in the stockpot, add the chopped vegetables and sauté gently for 5 minutes. Add the spinach, coriander, parsley and mint leaves, plus the chillies, and cook for another 3–5 minutes.

Pour in the stock and bring it to the boil. Add the rinsed quinoa and spring onions. Reduce the heat to a gentle simmer, then half-cover the pot and simmer for 30 minutes. Add some hot water to the soup if it is reducing rapidly.

Meanwhile, remove the chicken meat from the bones and cut into bite-sized pieces. When the soup has been cooking for half an hour, stir in the chicken and allow it to heat through. Season to taste with salt and pepper before serving.

Persian-style barley soup

Actually, this type of barley soup is not strictly Persian, merely popular in Tehran, according to Margaret Shaida who lived in Iran for 25 years and learnt to cook its traditional dishes from her Persian mother-in-law. In her book *The Legendary Cuisine of Persia*, Margaret theorizes that this soup entered Iran only in the early part of the 20th century, thanks to the White Russians fleeing the Bolshevik Revolution. This version differs from hers only in so far as it incorporates my own (often lazy) culinary habits and aversion to grating.

ingredients

2 tbsp olive oil

2 onions, finely chopped

2 leeks, finely chopped

2 carrots, finely chopped

150g/¾ cup semi-pearled or pot barley

1.5 litres/6¼ cups lamb stock

juice of 2 lemons

2 tbsp soured cream

a handful of chopped parsley

salt and pepper

serves 4

method

In a large, heavy saucepan, heat the olive oil and gently fry the onion until translucent. Add the leeks and carrots and continue frying gently until softened.

Add the barley and stock and bring to the boil. Half-cover the pan, reduce the heat and simmer for 1½–2 hours, stirring occasionally to prevent sticking, until the mixture is thick and the ingredients are breaking down. Add more stock or some hot water from the kettle if it reduces too quickly.

Pour half the lemon juice into the saucepan and continue cooking until the barley is soft. Remove from the heat and stir in the soured cream and most of the parsley, adding the remaining lemon juice to taste. Season to taste with salt and pepper, and serve sprinkled with the reserved parsley.

cook's notes

● Margaret's recipe uses pearl barley, and slightly more of it. I prefer the taste of pot barley and, wanting to keep only one variety in the cupboard, would use it in preference.

● The Persian way of incorporating the carrot is to grate it and add it along with the lemon juice, after the soup has been cooking for some time.

● Barley goes particularly well with lamb stock, but there is no reason why beef, chicken or a richly flavoured vegetable stock could not be used instead.

● The wheat family of grains make a good alternative to barley in this soup.

● For a barley soup in the German style, sauté some leeks and celery in butter, then add about half the quantity of barley used here and your choice of chicken or veal stock. Skip the lemon juice and soured cream, seasoning instead with nutmeg. To thicken the soup, make a 'liaison' of 120ml/½ cup cream beaten together with an egg yolk and slowly stir the hot soup into it. A garnish of parsley is optional.

Corn stock

I love the frugality of this recipe. It follows the admirable tradition of many of the world's poorer countries of making effective use of every part of the plant, or animal. Mexicans certainly put all parts of the maize plant to work, though not in this particular way. Simmered long enough, leftover corn cobs produce an interesting, intensely flavoured alternative to shellfish stock. When cooked for a shorter time they make a good light vegetable stock, adding welcome diversity to the range of tastes available to vegetarians for soups, sauces and stews.

ingredients

4–6 leftover corn cobs (kernels removed)

1.6 litres/7 cups water

1 small onion, halved

1 carrot, scraped

½ stalk celery

10 peppercorns

1–2 bay leaves

2 sprigs thyme

salt

makes 600ml/2½ cups

method

Using a cleaver or large cook's knife, chop each corn cob into 4 pieces. If you don't manage to cut right the way through the cob, don't worry; once a cut is made, the cob will easily snap into two.

Place the chopped cobs in a stockpot with the water. Add the onion, carrot, celery, peppercorns, bay, thyme and a little salt, and bring to the boil. Half-cover the pot, reduce the heat and simmer very gently for 2 hours.

Strain the stock, pressing down on the solids to extract as much flavour as possible. Use as is or place in a clean saucepan and boil rapidly over a high heat until the stock has reduced and its flavour has intensified to your liking.

cook's notes

● The recipe here might be called traditional in terms of its flavourings, but variations of it are in no way restricted to these aromatics, or indeed this method.

● Douglas Rodriguez (see page 101) includes leeks, saffron, jalapeno chillies and tomato paste in his version, sweating the vegetables and saffron in butter before adding the water and other herbs and spices.

● Joseph Sponzo (see page 106) uses some of the kernels from the corn cobs and includes only basil stems and tarragon sprigs as flavourings. He adds the herbs once the stock has been reduced and removed from the heat, and discards them before storing or further cooking. This infusion method of flavouring is favoured by some meticulous chefs who argue that boiling herbs over an extended period causes bitterness in the finished stock.

Hazelnut and feta salad with sprouts

Really good sprout salads are a rare thing. I've looked around but eaten only one worth writing about – an elegant little pile of virtue prepared by chef Stephen Bull several years ago when he had a restaurant in the City of London. This dish is based on it. Success lies not in the sprouts but the ingredients carefully chosen to match them. You don't often see cheese in a sprout salad, but as long as a little of what we fancy does us good, and helps the sprouts to do us good, what's the problem? Buy ewes' milk feta if you have a problem with cows' milk.

ingredients

150g/1 cup hazelnuts

2 oranges

5cm/2in cucumber, finely chopped

1–2 sticks celery, finely chopped

75g/2 cups buckwheat sprouts

75g/2 cups short mung bean or lentil sprouts

20g/1 cup alfalfa sprouts

150g/5oz feta cheese, crumbled

2 tbsp hazelnut oil

1 tsp Dijon mustard

flat-leaf parsley or salad cress, to garnish

salt and pepper

serves 4

method

Heat the oven to 200°C/400°F/Gas 6. Place the hazelnuts on a small baking tray and roast for about 20 minutes, shaking the pan frequently, until the skins are blackened. Remove from the oven and wrap the nuts in a clean tea towel. Set aside to cool.

Remove the peel and pith from one of the oranges. Holding it over a mixing bowl, use a serrated fruit knife to cut the segments of orange flesh from the connective tissue and let them fall into the bowl, collecting as much of the juice as possible.

Toss the chopped cucumber and celery into the bowl and set aside. Use the tea towel to help rub the charred skins away from the roasted hazelnuts. Discard the skins and chop the nuts in half, or more roughly if preferred. Add to the bowl with the sprouts and crumbled cheese, and toss well.

Juice the remaining orange and place in a small bowl with the nut oil and mustard. Whisk together, then season to taste with salt and pepper.

Stir the dressing into the salad and adjust the seasoning as necessary (the feta will be very salty). Serve on plates garnished with small sprigs of flat-leaf parsley or a few strands of cress, as desired.

cook's notes

● Despite what some health faddists may think, there is such a thing as too many sprouts. The quantities here are just a starting point. Adjust the amount of each ingredient to suit your own taste. You want to feel the salad is so delicious that you'd like to make it again one day, not that if you took just one more mouthful, you'd turn into a horse.

● Large Chinese-style mung beansprouts are not ideal but are an acceptable inclusion. Before adding them to the salad, blanch for 1 minute in a large pan of boiling water, refresh under cold running water and dry thoroughly.

● This is a complete meal in itself, a fine lunchbox dish or a good winter side salad for plainly grilled lamb or chicken (see page 198).

Wholemeal pasta

It's no wonder many people turn their noses up at wholemeal pasta. For many years the readily available commercial varieties made wholeness, rather than good quality pasta, their priority. Unsuitable flours were used, resulting in mushy strands with small flakes of bran that made them seem like wet shreds of corrugated cardboard. Thankfully, things have changed tremendously, and there are now several Italian companies producing excellent wholemeal pasta: La Terra e il Cielo is my favourite. Making your own is fun and easy, too.

ingredients

300g/2 cups fine wholemeal wheat-type flour

3 eggs

3 tablespoons water

serves 4

method

Place the flour in a large mixing bowl and make a well in the centre. Break the eggs into the well and break them up lightly with a fork or little whisk.

Gradually work the flour into the eggs, adding the water as you go, until the mixture combines to give a dough. Knead lightly in the bowl for a few minutes.

Dust a work surface with flour and transfer the dough to it. Knead for a few more minutes to give a smooth ball. Wrap in plastic wrap and leave to rest for 30–60 minutes or so before rolling.

Use a hand-cranked or electric pasta machine to roll out the dough, following the manufacturer's instructions. To cook, bring a large pan of salted water to the boil, add the pasta and cook for 1–2 minutes before draining.

cook's notes

● It's not just the variety of grain used for pasta flour that is important but also the degree of the grind. You need a very fine, powdery flour for best results. By far the best wholemeal wheat noodles I have made employed an organic farro flour or *farina integrale di farro* from a company called Energia dalla Natura.

● Scaling this recipe up or down is easy: 100g/⅔ cup flour, 1 egg and a tablespoon of water per person will give you a generous serving with a little left over.

● You can use a floured worksurface and rolling pin to roll the dough, but work in batches. After you have rolled out each batch thinly, cut it into thin strips using a large cook's knife, and set the noodles aside on a tray, keeping them covered to prevent drying.

● This basic recipe works well with finely milled freekeh, buckwheat and quinoa (if you like the flavour). If preferred, use half wheat and half of the other flour – a good option is rye, which tends to be too strong to use on its own. You could also use triticate, the wheat-rye hybrid, if available.

● Some people like to use 1 tablespoon of oil in place of the water. In theory, this gives a softer dough – something I'm not convinced you'll want.

Spaghetti with tomatoes, olives and walnuts

On a hot summer's day, this cold, elegant spaghetti dish is wonderfully refreshing, and a million miles closer to Italy than the anglicized pasta salads that have given cold pasta such a bad reputation. As with all wholemeal dough products, whether made by you or at a factory, texture is the key to pleasurable eating. A very thin cut of pasta such as spaghetti, angel hair or linguine best suits hard, nutty wholemeal dough and is essential here to the succulence of the finished dish. Wider noodles set you on the slippery road to Wet Cardboard Land.

ingredients

30g/ $\frac{1}{4}$ cup walnuts, broken

1 medium ripe tomato

100g/3 $\frac{1}{2}$oz wholemeal wheat-type linguine
    or spaghetti

7 marinated purple olives such as kalamata, stoned

1 tbsp olive oil or oil from the marinated olives

a small handful of basil leaves

salt and pepper

per person

cook's notes

● If you have been wondering how best to use those wholemeal spelt noodles in the health food shop, this is the dish to try. Simply scale this recipe up according to the number of people you are feeding. It works very well as a quick summer supper, lunch or starter.

● Use buckwheat noodles if preferred – the flavour is wonderful with juicy tomatoes – and add a little chilli and garlic to the sauce if desired.

● Like many people, I usually ignore instructions in cookbooks to skin tomatoes, but here the luscious effect of the peeled flesh is important. Given the lack of other cooking required for the dish, it does not greatly affect the preparation time or degree of effort, but the results are well worth it.

● Purple olives are specified merely because their colour enhances the dish. Use black olives if you have them to hand. If your olives are not already marinated in a lovely mixture of oil, herbs and an acidic element such as vinegar or lemon, add these ingredients to the finished dish yourself, adjusting the flavours to taste.

method

Put a kettle of water on to boil. Meanwhile, in a dry frying pan, lightly toast the walnuts until just fragrant, stirring constantly. Transfer to a bowl to cool.

With a small, sharp knife, score a cross in the rounded end of the tomato and place in a heatproof bowl. Pour some boiling water from the kettle over the tomato and leave for 1–2 minutes.

Pour the rest of the water from the kettle into a large saucepan. Add a little salt and bring to the boil. Cook the pasta according to the packet instructions.

Meanwhile, drain the hot water from the tomato and peel away the skin. Remove the core, then quarter the tomato and remove and discard the seeds. Dice the flesh and add to the bowl with the walnuts.

Roughly chop the stoned olives and place with the tomato and nuts. When the pasta is only just al dente, drain and refresh it under cold running water. When it has thoroughly drained, toss it with the tomato, nuts and olives, adding the olive oil or some of the leftover marinating oil from the olives.

Tear the basil leaves into the pasta, season to taste with salt and pepper and then toss again before serving at room temperature.

Spaghetti with sardines and fennel

This southern Italian dish was a favourite of London-based chef David Eyre while working at The Eagle, a renowned pub he co-founded in 1991 that serves good food and which is often credited with launching Britain's still-thriving gastropub movement. (David, in fact, was inspired by Franco Taruschio, the Italian former owner of The Walnut Tree Inn near Abergavenny in Wales.) Fresh sardines have traditionally been popular only in Mediterranean coastal regions but they were the first fish to be canned, which resulted in wider distribution.

ingredients

4 fillets of large fresh sardines, or 6 smaller ones

200g/7oz wholemeal wheat-type spaghetti

1 tbsp olive oil, plus extra to taste

1/2 fennel bulb, finely chopped

1 clove garlic, chopped

2-3 tbsp chopped flat leaf parsley

juice of 1/2 lemon

salt and pepper

serves 2

method

Pat dry the sardine fillets and sprinkle them lightly with salt. Bring a large saucepan of salted water to the boil, add the pasta and cook according to the packet instructions.

Meanwhile, heat the olive oil in a large frying pan or skillet, add the fennel and garlic and cook, stirring, for 30 seconds or until fragrant. Add the sardine fillets to the pan skin-side down. Cook for 1 minute, then turn them over and cook for another minute or until done. Remove from the heat.

Drain the pasta and return it to the saucepan. Add the sardine and fennel mixture, plus the parsley, and toss, adding a little extra olive oil if desired. Season to taste with salt, pepper and lots of lemon juice.

cook's notes

● David's recipe differs from mine in that it doubles the quantity of sardines per person and uses more olive oil. I use more fennel instead. Even though I'm not a great fan of its aniseed flavour, it does work well in this context with lemon and parsley, and increasing vegetable intake is a priority for many people these days.

● As with the dish on the previous page, this recipe is at its best with a very thin wholewheat pasta shape such as spaghetti or linguine.

● To clean, gut and fillet sardines yourself, wipe their scales away underneath cold running water and dry with kitchen towel. Press your thumb right along their backs to loosen the flesh from the spine. Using a small sharp knife, cut off and discard their heads, then cut lengthways along the bellies. Open the belly flaps and pull out and discard the innards. Press the fish open while running your thumb along one side of the spine to assist the flattening. Working from the neck ends, pull the spines out of the fish and snap them off at the tails. Separate the fillets, then rinse them clean and pat dry.

● The combination of wholewheat pasta and oily fish is not exclusive to southern Italy. In Venice, a traditional dish of wholewheat pasta features a sweet-sour sauce of anchovies and onions. The bigoli pasta is a long noodle but very thick, a point which may contribute to the dish's lack of popularity outside its local area.

Barley grit couscous

Couscous can be made authentically with many grains, and it especially suits low or no-gluten varieties. The North African Berbers do indeed make barley grit couscous, however the stew cooked underneath it here is not an authentic recipe. It has been chosen as an unintimidating introduction to the process of steaming barley couscous, something which (unlike rolling your own couscous from wheat semolina) you might tackle on a busy weeknight. Couscous is also made with corn, millet, green wheat and sorghum. You could also use cracked freekeh.

ingredients

150g/¾ cup barley grits, or uncracked semi-pearled
    or pot barley

2 tbsp olive oil

1 onion, sliced or chopped

1 clove garlic, chopped or crushed

400g/14oz canned chopped tomatoes

a large pinch of powdered saffron

600ml/2¾ cups water

1 tsp butter, plus extra for greasing

200g/7oz firm fish such as marlin, cubed

a large handful of basil, torn

salt and pepper

serves 2

cook's notes

● To make this a vegetarian dish, replace the fish with cauliflower florets cut small and add to the sauce just 2–3 minutes before serving.

● Millet couscous pellets, typical of African cooking, are easy to make, but freshly ground flour is essential otherwise the dish will have an unpleasant bitter aftertaste. You simply place the millet flour in a bowl and gradually work in water, rubbing the flour together with your fingers and raking through it until tiny balls are achieved.

method

If using uncracked barley, place it in the bowl of the grinding attachment of a food processor and pulse until the grains are cracked but not finely ground. Place the barley grits in a large bowl and cover with water. Set aside to soak.

Meanwhile, heat the oil in the base of the couscousiere or steamer, add the onion and garlic and fry gently until tender. Add the tomatoes, saffron and water and bring to the boil, then reduce the heat to low.

Arrange the top part of a couscousiere, or a steamer basket lined with buttered muslin, over the pan. Squeeze the excess water from the grits and add them gradually to the top section, rubbing them between your palms as you drop them in. Cover and leave to cook for 20 minutes.

Remove the top section of the pot from the heat and dump the grits into a large bowl or roasting tray. Break the lumps up with a fork and stir in the butter. Gradually sprinkle 6 tbsp of water over the grain, raking it in with the fork.

Stir the tomato sauce well, then return the barley to the steamer and place over the pan. Steam for a further 20 minutes. Remove the grain from the steamer and fork through again.

When the tomato sauce has thickened considerably, add the fish, basil and salt and pepper. Return the grain to the steamer, place back over the stew and steam for a further 10–15 minutes, or until the fish is cooked and the barley grits are tender.

Fluff up the barley again and place in warm serving bowls. Adjust the seasoning of the fish stew to taste, spoon it over the barley and serve.

Pap tart

This superb recipe was given to me by South Africa's highly esteemed restaurateur and food writer Peter Veldsman, a specialist in Cape cuisine and owner of Emily's restaurant, based in Cape Town's Waterfront area. It's a fascinating alternative to Italian recipes for layered and baked polenta, being very similar and yet noticeably different in its use of cinnamon, cumin and cayenne. So moreish is it that my boyfriend and I gorged ourselves on it all evening the first time we made it. It features in Peter's excellent *Flavours of South Africa*.

ingredients

350g/2 cups coarse cornmeal

2 tsp salt

1 tsp cumin seeds

1/2 tsp cayenne pepper

1 litre/4 1/2 cups water

2 tbsp butter

100g/3 1/2 oz fontina or parmesan cheese, sliced

For the sauce:

2 tbsp olive or vegetable oil

2 large onions, sliced

1 stick cinnamon

1 tsp cumin seeds

3 cloves garlic, crushed

2 tsp salt

freshly ground black pepper

225ml/1 cup water

1.2kg/2lb 10oz canned chopped tomatoes in juice

225ml/1 cup red wine

cayenne pepper, to taste

a large handful of basil leaves, torn

2 tsp chopped oregano or thyme

serves 6–8, in theory

method

Place the cornmeal, salt, cumin and cayenne in a large, heavy saucepan. Slowly whisk in the water and bring the mixture to a boil, stirring constantly. Lower the heat and simmer gently, stirring often, for 30 minutes. Stir in the butter, remove the pan from the heat and set aside.

Meanwhile, to make the sauce, heat the oil in a large saucepan and fry the onions, cinnamon and cumin until the onions are soft and translucent. Stir in the garlic, salt and pepper and water, and bring to the boil. Add the tomatoes and wine, and simmer for 30–40 minutes until the mixture is very thick, stirring occasionally. Season to taste, adding the cayenne and fresh herbs. Discard the cinnamon.

Line a 20cm/8in square baking tin with foil so that it hangs over the rim. Press a layer of polenta into the base of the tin, smoothing over the surface. Cover with a layer of tomato sauce, then repeat the layers to fill the tin, ending with polenta. Chill, covered, for 1 hour to set.

Heat the oven to 170°C/325°F/Gas 3. Lay the sliced cheese over the top of the tart and bake for 20–25 minutes or until piping hot. Serve.

cook's notes

● Peter recommends serving the finished dish scattered with chopped chillies, fried garlic cloves and fresh coriander or basil leaves. Fried slices of aubergine (eggplant) and grilled or barbecued meats are good accompaniments.

● Using red wine really lifts the tomato sauce, which is known as sheba in South Africa. You might want to include some chopped lavender in it, too – another idea of Peter's.

Soft, creamy polenta

Italian food experts of my acquaintance would insist that this is not polenta but a cream of maize. My choice of cheese would be taleggio, partly because I adore its fruity, herby, rather vegetal flavour, especially when just stirred into the hot cornmeal, so that the creamy paste only slightly melts and the rind stays a little crusty. The other reason I choose it is that Stefano Cavallini's excellent Italian deli-traiteur I Sapori is based near my London flat, and his taleggio is superb. You can find similar cheese at esperya.com.

ingredients

180g/1 cup coarse cornmeal
1 tsp salt
1 litre/4¹/₂ cups water
150–200g/5–7oz semi-soft Italian cheese, cubed
50g/¹/₄ cup butter, cubed
freshly ground black pepper
serves 4–6

method

Place the cornmeal and salt in a large, heavy saucepan. Slowly whisk in the water and bring the mixture to a boil, stirring constantly. Reduce the heat right down and simmer gently, stirring often, for 35 minutes. Alternatively, transfer the mixture to an oven heated to 180°C/350°F/Gas 4 and bake for 40 minutes, ensuring you use an ovenproof pan.

Remove the pan from the heat and beat in the butter. Stir in the cubed cheese and season to taste with salt and freshly ground black pepper. Serve hot.

cook's notes

● Try polenta taragna, a mixture of buckwheat and corn, instead of plain cornmeal. The Moretti brand is good, but you can make your own version by combining 100g/½ cup buckwheat flour (a dark, coarse grind is preferable) with 200g/1 cup of cornmeal.

● Other good additions to creamy polenta are chopped olives, sun-dried tomatoes, ham, piquant cured sausages and robust Mediterranean herbs, such as oregano and sage. Mix and match them, using parmesan or pecorino cheese and/or olive oil as desired, to suit your own preferences. Remember to add plenty of black pepper.

● Walnuts and blue cheese is a particularly good combination in a dish such as this.

● The secret of the most delectable soft polenta often is, unfortunately, nothing more than beating in a very large pack of butter at the end of cooking. Skip the cheese on these occasions because you don't want to run the risk of adding anything remotely healthy like protein to this mixture. Top your buttery polenta with some fresh porcini, sliced and sautéed, with a little garlic, in lots of butter. It's a chilly weather dish, so you can wear big baggy jumpers for a week or two afterwards without looking too foolish.

● You can use any of these ideas with grits for a modern spin on Southwestern USA fare.

Sweetcorn risotto

Mixed grain risottos are proving a fun area of experimentation for many chefs cooking in a contemporary style. Almost anything goes, depending on your tastes, but retaining the white rice is key when coaxing anyone wary of unusual grains into trying such a dish. Here the surprising use of sweetcorn adds a pleasing fresh chewiness that appeals to many people who don't like traditional risotto. This recipe is similar to one enjoyed at Bobby Flay's sensational Mesa Grill in New York and was served at my engagement party with much success.

ingredients

850ml/3¾ cups rich vegetable or chicken stock

2 large ears sweetcorn, about 16cm/6½in long,
*   husked and strings removed*

6 spring onions

50g/¼ cup butter

1 tbsp olive oil

300g/1½ cups risotto rice

150ml/⅔ cup white wine

50–75g/½–¾ cup parmesan cheese, grated,
*   plus extra to serve*

salt and pepper

serves 4

method

Put the stock into a small saucepan and bring slowly to the boil. Meanwhile, remove the kernels from the sweetcorn and place them in a bowl with the creamy juices they exude. Break up any clusters of kernels, then set aside. Chop the spring onions, keeping the white and green parts separate.

In a large, heavy saucepan, melt half the butter with the oil over a moderate heat. Add the white parts of the spring onions and sauté for 5 minutes or until soft and tender but not browned.

Add the rice and cook for a few minutes, stirring, until the grains are shiny and becoming translucent. Pour in the wine and keep stirring the rice until the liquid bubbles away.

Add a ladle of the simmering stock and make a note of the time. Cook, stirring constantly, until the rice has absorbed all the stock. Then add another ladle of hot stock and repeat the process.

Once you have been adding stock and stirring the risotto for 15 minutes, add the corn kernels, their juices and 4 heaped tbsp of the reserved green parts of the spring onions. Continue adding stock and stirring the risotto for 5 minutes.

Turn off the heat and stir in the cheese and the remaining butter. Season to taste with salt and pepper. Sprinkle with a little of the leftover spring onions and serve immediately with extra cheese.

cook's notes

● Authenticity is not an issue here, so do not think Parmigiano Reggiano is essential. Substitute any well-flavoured grating cheese such as cheddar, pecorino, manchego or, given the American inspiration, Dry Jack.

● If you want to add herbs, think basil, coriander or parsley. A little red or green chilli works here, too.

● The stripped corn cobs can be used to make corn stock (see page 103). Using corn stock in this dish gives a paler and more delicately flavoured result than rich vegetable or chicken stock do.

● The sky will not fall in if you make up some powdered vegetable bouillon using fresh boiling water and add it from a jug at the side of the stove rather than from a simmering pan. Note that most chefs have also risked hell-fire and damnation by not stirring risotto constantly for 20 minutes.

● Add protein in the form of crisp-cooked bacon crumbled over the top (three rashers for four people), grilled prawns (see page 198), marinated beef kebabs (page 198), or some slices of barbecued, grilled or roasted chicken or turkey. You could top each serving with a whole roasted or deep-fried chilli, mild, of course.

● This mixture, or any leftovers, can be made into burgers or arrancini by adding a little flour and some beaten egg. Shape into balls or patties, then dust with cornmeal or breadcrumbs. Shallow or deep-fry in hot oil until the coating is golden brown.

● Paul Gayler of The Lanesborough makes a mixed grain risotto by adding part-boiled barley and quinoa to the rice towards the end of cooking.

Barley 'risotto'

The lovely photograph opposite was taken at Luke Mangan's restaurant Salt in Sydney. He's just one of several chefs to have experimented with orzotto, as it is sometimes termed (*orzo* means barley in Italian), but he takes the trophy for most creative use of it: surrounding the grains with coconut broth and topping with roast barramundi. When asked why he thought the unusual combination would work he confessed that some free-flowing alcohol while jetting to America may have been a factor in its inception, yet it does work, and beautifully.

ingredients

1.25 litres/5 cups fish or chicken stock

2 tbsp butter

1 tbsp olive or vegetable oil

2 shallots, finely chopped

200g/1 cup pearled or semi-pearled barley

150ml/^{2}/$_3$ cup white wine

4 tbsp cream or coconut cream, or to taste

a handful of chopped flat-leaf parsley

salt and pepper

serves 4

method

Put the stock into a saucepan and bring slowly to the boil. Meanwhile, in a large, heavy saucepan, melt the butter and oil over a moderate heat. Add the shallots and sauté for 5 minutes or until soft.

Add the barley and cook for 3 minutes, stirring, until the grains are shiny. Pour in the wine and stir until it bubbles away. Add a ladle of hot stock. Cook, stirring constantly, until the barley has absorbed all the stock. Then add another ladle of hot stock and repeat the process. Continue until the barley is tender, adding hot water as necessary once all the stock has been incorporated.

Stir in the cream or coconut cream and the parsley. Season to taste and serve.

cook's notes

● First know your barley. White, polished pearl barley is normally favoured for this type of dish because it releases its starch readily, yet it can be difficult to buy. The slightly beige semi-pearled barley is nuttier, chewier, and requires longer cooking, but is still very pleasant, if not better.

● If you are using semi-pearled barley, start adding hot water after this measure of stock has been incorporated, otherwise the dish will taste too strongly of reduced stock.

● An alternative method of cooking is to boil semi-pearled or pot barley in water for about 30 minutes, then drain and cook as for traditional risotto. Or you could simply cook the barley until tender in stock, drain and sauté with some cream or mascarpone and chopped herbs.

● This last technique is used by French chef Emmanual Leblay for making a 'risotto' of *petit épautre*, featured in Patricia Wells' *Paris Cookbook*. Semi-pearled farro can also be used in this fashion. Served with roast chicken and a sauce of warmed reduced chicken stock, it was one of the best grain dishes I have ever eaten, and remarkably easy.

● For pan-roast fish and coconut broth recipes, see pages 198 and 201.

Genmai gomoku

The Japanese do a lot more with rice than make sushi. Gomoku dishes, in which a variety of tasty morsels are simmered with rice in a pot, are a good example. The Japanese don't, however, as a rule like brown rice, which this version features. It is an excellent way to use all those delicious plump-grained brown rices that Europeans won't touch either but love to export. Women will be particularly gratified to find such a moreish way of consuming those intriguing packs of snow-dried tofu and seaweed found in fashionable organic stores.

ingredients

4 squares snow-dried tofu

5g/ $^1/_4$ cup hijiki seaweed

400g/2 cups short-grain brown rice

850ml/3$^3/_4$ cups dashi, fish or vegetable stock

4 tbsp soy sauce

2 tbsp mirin

4 fresh shiitake mushrooms, destemmed
   and sliced

1 small carrot, sliced

1 small turnip or daikon radish, diced

serves 4

cook's notes

● If you would like to add some fresh greenery to this dish, add 12 trimmed mangetout or sugar snap peas to the pot a few minutes before the rice is scheduled to finish cooking.

● Dried shiitake can be used instead of fresh, in which case the flavoured soaking water can be strained and used in place of some of the stock. Remember to discard the tough stems.

● This is a good opportunity to practise your carrot carving skills; it really enhances the look of the dish. To do this, cut 4 or 5 equidistant little grooves down the length of the peeled carrot before slicing it.

method

Place the tofu in a small heatproof bowl. Cover with warm water and leave to soak for 5 minutes. Squeeze the tofu to release the milky liquid, then discard the water and soak again. Repeat the process 4–5 times or until the water that oozes from the tofu is no longer milky. Meanwhile, soak the hijiki according to the packet instructions.

Rinse the rice and place in a large, heavy pot or casserole. Add the stock, soy sauce and mirin, cover and bring to the boil. Reduce the heat and simmer for 25 minutes.

Using scissors, cut the squares of drained tofu into quarters. Cut any large pieces of hijiki. When the rice has been cooking for 25 minutes, lift the lid and scatter in the tofu, seaweed, mushrooms, carrot and turnip or daikon. Cover again and continue cooking for 15 minutes until the liquid is absorbed and the ingredients are tender.

Turn off the heat and leave the pot to stand, covered, for 10 minutes. Lift the lid, stir the rice very gently and spoon into warm bowls to serve.

Lovely lamb stew

Stew is not one of the English language's most enticing words and, as a result, many wonderful traditional dishes are being ignored. Revisiting the old-fashioned British lamb stew was one of the great delights of compiling this book. The preparation is almost mindless and yet with long, slow cooking the ingredients meld into a creation that tastes as though it took a clever chef a lot of effort to produce. There is no need for side dishes or tricksy presentation. All that's required to make it look sumptuous is a variety of vegetables.

ingredients

100g/ ½ cup semi-pearled or pot barley

2 tbsp oil

950g/2lb 2oz lamb chops with bones,
    such as those from the neck end

2 small parsnips, cubed

1 medium onion, chopped

1 large leek, thickly sliced

250g/9oz carrots, cubed

250g/9oz swede, cubed

4 small potatoes, quartered

1 bouquet garni, made by tying together a small
    length of celery, a bay leaf, a few sprigs of thyme
    and some parsley stalks

some water, stock, ale or cider, to cover

a handful of chopped parsley

salt and pepper

serves 4–6

method

Put the barley in a saucepan, cover generously with water and bring to the boil. Reduce the heat and simmer for 15–20 minutes to part-cook it.

Meanwhile, in a large flameproof casserole, heat the oil. Working in batches, add the lamb and brown the meat on each side. Remove to a plate and add the parsnips, onion, leek, carrots and swede to the pot and cook, stirring occasionally, for 3 minutes until they begin to soften. Scoop any excess fat from the pan. Preheat the oven to 180°C/ 350°F/Gas 4.

Drain the barley and add it to the pot with the browned meat, potatoes and bouquet garni. Season with some salt and pepper, then pour in just enough water, stock, ale or cider to come about 1cm/½in under the top of the ingredients in the pot. Cover and bring to the boil, then transfer the casserole to the oven for 2 hours.

About 15 minutes before the end of cooking, stir the chopped parsley into the casserole. Remove the bouquet garni and season to taste before serving.

cook's notes

● Terrific though barley is, this dish also works well with short-grain brown rice and wheatberries.

● In his book *Appetite*, Nigel Slater suggests that stews such as this one are good made with duck or guinea fowl as well as more popular chicken.

● If preferred, replace the root vegetables in this dish with canned tomatoes, a little celery, a bell pepper and some black or green olives. Add less liquid (you could use wine) to compensate for the juiciness of the tomatoes.

Mushroom ragout with barley

According to the *Larousse Gastronomique*, the French word *ragout* dates from 1642 and was once used to describe anything that stimulated the appetite. The picture opposite certainly does that for me. Today ragout is understood to mean a stew made from meat, poultry, game, fish or vegetables cooked in a thickened flavoured liquid. However, ragouts of vegetables and mushrooms are usually cooked in their own juices after first browning the ingredients. This hearty recipe comes from Poland, where wild mushrooms are rather commonplace.

ingredients

15g/ ¹⁄₂oz dried morels

15g/ ¹⁄₂oz dried porcini

700ml/3 cups chicken or vegetable stock

2 tbsp butter

1 large onion, finely chopped

1 stalk celery, finely diced

100g/ ¹⁄₂ cup semi-pearled or pot barley

1 bay leaf

¹⁄₂ tsp salt

300g/10¹⁄₂oz fresh mixed selection of exotic
  and wild mushrooms, preferably including chanterelles

125g/ ¹⁄₂ cup soured cream or crème fraîche

a handful of fresh parsley, roughly chopped

serves 4

cook's notes

● This looks better and is more enjoyable to eat if the cool cream is folded in just a little so that there are pockets of creaminess, rather than stirring until it is thoroughly combined. Alternatively, serve the fresh sautéed mushrooms on a bed of the barley stew and use the cream more like a sauce, drizzling some over the mushrooms and stirring some into the ragout.

● Wild rice, wholegrain rice and wheatberries are good alternatives to barley in this dish. A portion of rye could be mixed into the barley too.

method

Pick over and clean the dried morels and porcini as necessary. Place the stock in a large saucepan and bring to the boil. Turn off the heat and add the dried mushrooms. Set aside to infuse for 30 minutes.

Melt half the butter in a frying pan and add the chopped onion and celery. Sauté for 2 minutes, then lower the heat right down, cover and cook gently for about 20 minutes until soft and sweet, stirring occasionally.

Using a slotted spoon, scoop the soaked mushrooms out of the stock and squeeze the excess liquid back into the saucepan. Roughly chop the rehydrated mushrooms and return them to the stock.

Add the sautéed onion and celery to the stock along with the barley, bay leaf and salt. Bring to the boil, cover and simmer over a very low heat for 45–50 minutes or until the barley is just tender and hardly any liquid remains in the saucepan. If the pan boils dry too quickly, top it up with a little hot water from the kettle.

Meanwhile, clean and trim the fresh mushrooms as necessary and, if large, cut into generous bite-sized pieces. Melt the remaining butter in a frying pan and sauté the mushrooms for 5 minutes until just cooked.

When the barley is ready, stir the sautéed mushrooms into it, then the soured cream or crème fraîche. Serve sprinkled with chopped parsley.

Posole verde

Authenticity of dishes is to be admired in food writing, but there's something about a recipe asking you to skin and bone a pig's head and a whole chicken that doesn't inspire most people to run to the kitchen. That's why I use this fruitily fragrant posole recipe based on one from Dean and Deluca's cookbook, rather than one from the great authorities on Mexican cuisine. The meat can be bought ready-prepared from your butcher. The only thing stopping me calling it a quick and easy dish is the fact that the corn grains need lengthy soaking and cooking.

ingredients

175g/1 cup red posole

175g/6oz green chillies

2 tbsp vegetable oil

1 large onion, finely sliced, plus 1 small onion,
    finely chopped, to garnish

2 cloves garlic, finely chopped

450g/1lb boned pork shoulder, cubed

4 chicken thigh fillets, quartered

50g/2½ cups fresh coriander, chopped

4 tbsp lime juice

1 avocado, diced

a small bowl of tortilla chips

salt and pepper

serves 6

cook's notes

● Hominy can be used in place of red posole, which I prefer for its stunning appearance but, at the time of writing, it is difficult to find.

● Don't skimp on the coriander and lime juice. Yes, it does seem a huge quantity but they make the stew taste wonderfully fresh and aromatic.

● Almost any green chilli will do. People from the Americas may balk at this idea, however, in other countries we usually have to make do with what we can get and in this case the differences in flavour do not present a huge problem.

method

Soak the posole in a bowl of cold water for 3 hours or as directed on the packet.

Meanwhile, roast the chillies over a naked flame or under a hot grill, until the skins have blistered and charred. Place the chillies in a plastic bag to steam the skins loose. When cool enough to handle, peel and deseed them. Chop the flesh coarsely and set aside.

In a large, heavy pot or casserole, heat the vegetable oil. Add the sliced onion and cook over a moderate heat, stirring often, for 8 minutes or until the onion is soft. Add the garlic and cook for another 2 minutes.

Add the pork, chicken and the roasted chillies and cook, stirring frequently, for 10 minutes or until the meat has started to brown.

Drain the posole and add it to the pot. Pour in enough hot water to just cover the meat. Bring to the boil, then reduce the heat right down, half-cover the pot, and cook the stew gently for 2 hours or until the posole is full-blown and the meat is meltingly tender. Add some more water if the liquid reduces too quickly.

Season to taste, then sprinkle with the chopped coriander and lime juice. Serve accompanied by separate bowls of diced avocado, chopped onion and tortilla chips.

Shellfish and saffron stew

Rich saffron sauce, thickened with eggs and cream, is a sensuous complement to the fresh shellfish and spiky black wild rice used in this luxurious dish. It's a real dinner party special that is as delectable visually as it is on the palate, yet relatively simple. Saffron is often held up as an expensive ingredient but it is tremendously good value as a little goes a very long way. Stored properly, the stamens last for years, unlike other spices. Despite what other recipes may say, it is important to toast and crush saffron before use to optimize its flavour.

ingredients

200g/1 cup wild rice

1/2 tsp saffron threads

5 egg yolks

500ml/2 cups cream

300ml/1 1/3 cups fish stock

300ml/1 1/3 cups dry white wine

750g/1lb 10oz green raw prawns, peeled
    and shells reserved

750g/1lb 10oz scallops

15 mussels, scrubbed

4 tbsp parsley leaves, chopped if desired

salt and pepper

serves 6–8

method

Cook the wild rice in a large saucepan with a generous quantity of boiling salted water.

Meanwhile, in a dry pan, toast the saffron for 2 minutes or until crisp and fragrant. Transfer to a mortar and crush until fine. Add 1 tbsp hot water to dilute the saffron.

In a large mixing bowl, beat together the egg yolks and cream, then add the saffron liquid and set aside to infuse for 30 minutes.

Drain and thoroughly rinse the wild rice when it is cooked, and set aside.

In a stockpot, combine the fish stock and wine and bring to the boil. Add the prawns and scallops and immediately remove from the heat. Allow to steep for 3 minutes, then use a slotted spoon to remove the shellfish from the stock.

Return the stock to simmering point. Add the mussels, cover the pot and cook until the mussels have opened, about 3–5 minutes. Remove them from the pan with a slotted spoon and set aside. Discard any that do not open.

Add the reserved prawn shells to the stockpot and bring it to the boil. Boil vigorously over a high heat for 20 minutes or until the stock has reduced to a volume of about 250ml/ 1 cup and has intensified in flavour.

Line a sieve with a double thickness of muslin and pour the stock through it into a large jug or bowl. Discard the solids.

Slowly whisk the hot stock into the saffron cream, then return the mixture to the pot. Add the cooked wild rice, prawns and scallops and heat through over a low heat. As soon as it is hot, add the mussels, turn off the heat, cover the pot and leave to stand for 1 minute so that the mussels just warm through.

Season the mixture to taste with salt and freshly ground black pepper. Ladle the stew into bowls, garnish with parsley and serve.

cook's notes

● Any wholegrain rice works well in this dish, as does black barley. If you rinse coloured rices after cooking, they will not leech too much into the cream sauce. An advantage of wild rice, though, is that its colour does not bleed at all.

● I've had equal success making this with large

pieces of firm white and pink fish in place of the scallops and prawns. In fact the recipe would work terrifically well with poached chicken, too.

● Include some vegetables if you like, especially sliced leeks and carrots, cooked briefly after you have removed the mussels from the pot.

Grünkohl

Pat Lawrence, husband of my food writer friend Sue Lawrence, has the dubious distinction of having been twice crowned Grünkohl Konig of his local village while living in Ostfriesland. The award is not for cooking the stew but for eating vast quantities of it. Grünkohl dinners often take place after inter-village matches of *bosseln*, a rowdy winter sport involving on-street bowling and a great deal of schnaps. Pat remembers them being 'like an Ostfriesich Burns night, but without the speeches and poetry'.

ingredients

2 tbsp vegetable oil

2 onions, chopped

1.25kg/2lb 12oz spring or collard greens

500g/1lb 2oz gammon or bacon joint,
*   presoaked if necessary*

500ml/2 cups water

3 tbsp coarse oatmeal or oat groats

2–4 wurst or other fresh sausages

salt and pepper

serves 4

method

Heat the oil in a large, heavy, flameproof casserole over a moderate heat. Add the onions, reduce the heat and cook gently for 10 minutes or until the onions are soft, stirring occasionally.

Meanwhile, trim the greens, separate the leaves from the stems and chop roughly. Add to the onions and cook, stirring frequently, for 5 minutes.

Lay the gammon or bacon joint in the casserole and pour in the water. Cover and cook over a very gentle heat, using a heat diffuser if necessary, for 1 hour, stirring occasionally to prevent sticking. Add more water if the pan becomes dry.

Add the oatmeal or groats, plus a little more water if necessary (this will depend on how gentle the heat is), then cover again and cook for 30 minutes, stirring occasionally.

Lay the sausages in the casserole and continue cooking for another 30 minutes.

Season to taste with pepper and, if necessary, some salt, although the mixture should already be rather salty, thanks to the cured meat. Serve hot, with boiled potatoes, beer and schnaps if desired.

cook's notes

● The first time I made this, the pan boiled dry near the end of cooking and some of the greens turned rather brown. It was still really yummy.

● Oats are a revelation in this type of stew, their creamy texture and sweet taste balancing the saltiness of the meat and the bitterness of the greens. But you could certainly use barley, wheat, wholegrain rice or rye.

● Despite living in England, where spring or collard greens are common, for more than a decade, this dish is the first to persuade me that I might purchase them regularly. However, a leafy savoy or other green cabbage will work fine.

● If you can't find wurst, go for for a mild herbed or spiced sausage. It doesn't really matter what type you choose. I have even used Toulouse sausages with great success in this recipe.

Chicken with freekeh

Allspice, also known as Jamaica pepper, makes a terrific general seasoning, especially for chicken dishes. The name refers to the taste, which is reminiscent of a blend of other spices such as nutmeg and cloves. Do not confuse it with products such as five-spice powder, which really are a blend of spices. The allspice berry looks like a large smooth peppercorn. Although native to the tropical areas of the Americas, it has become a distinguishing characteristic of Arab cooking. This recipe is based on one from Lebanese food expert Anissa Helou.

ingredients

1 chicken

300g/1½ cups freekeh

2 litres/8¾ cups water

3 sticks cinnamon

½ tsp allspice berries

6 tbsp butter (optional)

¼ tsp pepper

salt

serves 6–8

method

Rinse the chicken and place it in a heavy stockpot with the freekeh, water and cinnamon. Bring to the boil and skim any froth that develops on the surface of the water. Cover the pot and reduce the heat to medium–low. Cook the chicken for 1 hour, being sure not to let the water boil hard.

Meanwhile, toast the allspice berries in a dry frying pan until fragrant, stirring constantly. Transfer to a mortar and crush until fine.

When the chicken is done, use two large forks to lift it carefully from the cooking liquid. Make sure you tilt the cavity of the chicken downwards so that the hot liquid secreted inside runs back into the pot.

Place the chicken on a large baking tray and set aside to cool slightly. Remove the cinnamon sticks from the pot and discard, then turn the heat under the pot to low. Cover the pot and leave the freekeh to continue cooking gently while you remove and discard the skin from the bird. Cut the flesh from the bones and then into bite-sized pieces. Stir the chicken meat into the freekeh and continue cooking for 20 minutes, adding more water if the pan becomes too dry.

Meanwhile, if desired, melt the butter in a saucepan and cook until it is nut-brown. Stir the butter into the freekeh stew. Season with the allspice and pepper, adding salt to taste. Serve hot.

cook's notes

● This can also be made (and authentically so) with wheatberries or barley. In these cases the cooked chicken is traditionally cut into small pieces and beaten into the grain mixture so that the dish is more like a savoury porridge.

● According to Anissa, a shoulder of lamb, trimmed of fat, can be used in place of chicken.

● At the end of cooking, you could also stir in a handful each of parsley, coriander and mint, and a squeeze of lemon juice.

With buckwheat

With camargue red rice

With freekeh

With quinoa

cook's notes

● This basic recipe is easily varied. For a Middle Eastern flavour, replace the wild rice with freekeh, the dried cranberries with chopped dried apricots, the pecans with skinned almonds, pistachios or pine kernels (or a mixture), the Madeira with white wine, and add some ground cinnamon and allspice to the simmering onion and celery. Bulgur wheat will also work beautifully with these flavourings .

● If you like Lyn Hall's quinoa salad (page 105), replace the wild rice with quinoa (cooking it for only 12 minutes), the dried fruit with fresh grapes, the pecans with toasted cashews, the Madeira with white wine, and add some grated lime zest.

● Camargue red rice or other wholegrain red rice makes great visual impact. Instead of the fruit use 75g/3oz cubed pancetta fried along with the onions, and replace the nuts with blanched and peeled broad beans.

● Black rices can look stunning but beware of the colour leeching. That's one reason why I prefer wild rice.

● You could also replace the wild rice with buckwheat cooked as for kasha (page 150). Use 75g/3oz cubed bacon or pancetta instead of the fruit and 50g/2oz chopped fresh mushrooms instead of the nuts. Fry the meat and mushrooms along with the onion, include plenty of herbs and, if desired, some finely chopped spring onion.

● Rub the chicken with some crushed garlic if liked before roasting.

● An alternative method of cooking is to wrap the stuffing in a double thickness of kitchen foil and roast it alongside the chicken for the last 30–40 minutes.

Grain-based stuffings for chicken

A good independent butcher is a terrific asset to the cook, not just for careful sourcing of the best meats and poultry, but also for knife skills. They can perform much if not all of the fussy preparation that is essential for spectacular food presentations and, in today's competitive retail market, are usually keen to be asked to do so. De-boning whole chickens is a good example of how a butcher can do the complicated work for you.

ingredients

1 chicken, about 1.9–2kg/4lb 4oz–4lb 8oz, boned

butter, for smearing

For the stuffing:

200g/1 cup wild rice

1 tbsp olive or vegetable oil

1 small onion, finely chopped

1 stalk celery, finely chopped

4 tbsp Madeira

75g/ 1/2 cup dried cranberries or cherries

50g/ 1/2 cup roughly chopped pecans or walnuts

*1 tsp parsley leaves, or 1/2 tsp fresh thyme
   leaves, chopped*

salt and pepper

serves 6

Basic stuffing with wild rice

method

To make the stuffing, cook the wild rice in a saucepan of boiling salted water for 45–50 minutes. Meanwhile, in a frying pan, heat the oil and add the onion. Cook for 5 minutes, stirring often, then add the celery and continue cooking for 5 minutes until the onion is soft and golden. Add the Madeira and simmer, stirring occasionally, until it has evaporated.

Rinse the cooked wild rice thoroughly and drain well. Place it in a large bowl. Add the onion mixture, dried fruit, nuts, herbs and salt and pepper to taste.

Heat the oven to 220°C/425°F/Gas 7. Open out the boned chicken skin-side down on a work surface and pile the stuffing mixture into the centre, patting it together with your hands. Wrap the body back around the stuffing and use a large needle and kitchen string to sew the chicken back together, working from neck end right down the back of the bird to the tail.

Pat the chicken into shape and rub generously all over with butter. Season with salt and pepper. Place the chicken breast-side down in a roasting tin and roast for 20 minutes. Then reduce the heat to 190°C/375°F/Gas 5 and cook a further 20 minutes.

Remove the tin from the oven and turn the chicken onto its back. Baste well and return to the oven for 25–40 minutes or until the juices run clear when the thickest part of the thigh is pierced with a skewer.

Transfer the chicken to a carving plate and leave to stand for 10–15 minutes. Meanwhile, sieve the cooking juices from the tin into a jug and pour over the chicken when serving.

Freekeh pilaff

Green wheat freekeh's distinctive smoky taste, as well as its high levels of usable protein and fibre, make it an ideal base for vegetarian main courses. Adding nuts and fruit in typical Middle Eastern style enhances the flavour and increases the nutritional punch that this interesting grain packs. The delicious, luxurious recipe here is derived from one in Sonia Uvezian's *Recipes and Remembrances from an Eastern Mediterranean Kitchen*, in which she notes that freekeh is referred to in the Bible by the name 'parched corn'.

ingredients

60g/ $^{1}/_{2}$ cup pistachios

$^{1}/_{4}$ tsp allspice berries

3 tbsp olive oil

1 onion, finely chopped

200g/1 cup freekeh

570ml/2 $^{1}/_{2}$ cups vegetable stock

1 tbsp butter

50g/ $^{1}/_{3}$ cup dried apricots, sliced

6 tbsp raisins or sultanas

$^{1}/_{2}$ tsp ground cinnamon

salt and pepper

serves 4

cook's notes

● Following this basic recipe you can also make pilaffs based on bulgur wheat and rice. Reduce the quantity of liquid to 340ml/1½ cups and cut the cooking time to 15 minutes.

● Slivered almonds or pine kernels may be used instead of pistachios, but the colour of pistachios enhances the presentation. Alternatively use prunes, or a little chopped preserved lemon, instead of the dried apricots.

● Precede this dish with a selection of mezze based on fresh veg, such as green beans cooked in a sauce of tomato and onion, and a plate of crudités with herbs.

● This pilaff makes an excellent side dish for grilled chicken fillets (page 198) seasoned with a little crushed allspice, or roast lamb. You can also use it as a poultry stuffing by following the technique on page 143 using ready-boned chicken .

method

In a dry heavy frying pan, toast the pistachios over a medium to low heat, stirring constantly, until lightly browned and fragrant. Transfer to a bowl to cool. Repeat using the allspice berries, but transfer them once toasted to a mortar and crush finely.

In a large casserole, heat the olive oil. Add the onion and cook for 10 minutes or until soft and golden. Add the freekeh and cook, stirring for 2 minutes. Pour in the stock and bring the mixture to a boil. Lower the heat, cover the pan and simmer for 30 minutes or until the liquid is absorbed and the grains are tender.

Meanwhile, in the frying pan, gently melt the butter. Add the apricots, raisins, crushed allspice and ground cinnamon and cook, stirring frequently, over a medium-low heat until the fruit is plump and just starting to turn golden.

Roughly chop the toasted pistachios and stir them into the cooked freekeh. Stir in the sautéed fruit and adjust the seasoning to taste with salt and pepper. Serve hot.

Farro, pine kernel and sultana sauté

The scrumptious-sounding combination of wheat grains, pine kernels and sultanas goes back at least as far as Roman times. It features in Apicius, also known as *The Roman Cookery Book*, a work over 2,000 years old that is generally recognized as the first published collection of recipes. This sauté can be made and served without the chicken, however the flavour really benefits from being cooked in the same frying pan. The process of vigorously mixing the caramelized cooking juices with liquid to make a sauce is known as deglazing.

ingredients

1½ tbsp semi-pearled farro or other
   semi-pearled wheat-type grain

2 chicken breast fillets

2 tbsp olive oil

2 tbsp chicken or meat stock

4 tbsp pine kernels

4 tbsp sultanas

1 tbsp balsamic vinegar

salt and pepper

serves 2

method

Place the farro in a small saucepan with a generous quantity of salted water and bring to the boil. Reduce the heat and simmer for 25–35 minutes or until the farro is tender.

Meanwhile, place a chicken fillet under a sheet of plastic wrap and use a mallet or heavy rolling pin to flatten it evenly. Repeat with the other fillet.

When the farro is done, rinse it thoroughly and drain well. Heat the olive oil in a very large frying pan. Add the chicken and cook for 2–3 minutes on each side until cooked through. Remove from the pan and set aside in a warm place.

Pour the stock into the frying pan and stir vigorously with a wooden spoon to dissolve the caramelized cooking juices from the chicken. Add the farro, pine kernels and sultanas and sauté for 5 minutes until the pine kernels start to brown and the sultanas are plump.

Add the vinegar to the pan and toss to combine. When the sauce is thick, season the grain mixture to taste with salt and pepper and serve over the cooked chicken breasts.

cook's notes

● Farro expands a lot on cooking so unless you want leftovers, you raise the seemingly tiny quantity given here at your peril. There should be an equal volume of cooked grain (4 tbsp) to pine kernels and sultanas in the finished dish.

● Mix in a handful of chopped parsley towards the end of cooking if desired.

● One of chef Giorgio Locatelli's signature dishes is based on similar but more complex lines. He uses duck breasts, and they would certainly work here, but remember to spoon the vast quantities of fat they produce out of the pan before adding the grains and pine kernels.

● For a vegetarian dish, cook some baby savoy cabbages while the farro is boiling. Heat the oil in the frying pan and simply sauté the grains, pine kernels and sultanas in that, then mix in the vinegar and season. Serve the sauté over the cooked, drained cabbage, which you have cut into wedges and arranged on a plate.

● Crab apple or sour grape verjus (or verjuice), another modern and yet ancient ingredient, can be used in place of the balsamic vinegar for a lighter but still strongly fruity taste.

Herring in oatmeal

Herring is native to the North Atlantic but remains relatively unknown in the UK, despite it being rich in healthy fish oils. In his *New Delicatessen Food Handbook*, Glynn Christian theorizes that its lack of popularity in Britain may result from the fish's former association with poverty, but I suspect it's simply down to the bones. There's a lot of them, and for the inexperienced, eating your way round them takes a degree of practice and patience. Do not be concerned if you accidentally consume some: they are soft enough to eat and rich in calcium.

ingredients

50g/ ¼ cup medium or pinhead oatmeal

4 herrings, part-boned if desired

4 large rashers fatty bacon

a little vegetable oil or butter

a little mustard, to serve

salt and pepper

serves 2

method

Spread the oatmeal on a plate and season it with salt and pepper. Press the fish into the oatmeal, turning them to coat evenly on both sides.

Place the bacon in a large, heavy frying pan and set over a low heat. Slowly cook so that the fat from the bacon melts a little into the pan to provide the cooking medium for the fish. Cook until the bacon is a little browned, then remove from the pan and keep warm.

Place the coated fish in the pan and add a little oil or butter if there doesn't seem to be enough fat to fry the fish. Cook over a moderate heat for 3 minutes on each side or until the fish is cooked through. Serve with the bacon and a dab of mustard.

cook's notes

● The oatmeal should stick easily to fresh fish. If they're a bit older, place them briefly under cold running water and shake dry before coating in the oatmeal.

● The idea behind this traditional method of cooking herring is that the fish will be flavoured by the bacon fat. If you would rather cook the fish in oil and butter, that's fine. If you prefer, you can use herring fillets instead of whole fish, or trout instead of herring.

● The method of simply pressing the fish into the oatmeal works equally well with cornmeal. In this case you can keep the seasoning to just salt and pepper, or add a mixture of herbs and spices. For a 'blackened' flavour, combine 2 tbsp cornmeal with 3 small cloves of crushed garlic plus 1 tsp each of cayenne pepper, fennel seeds, dried oregano and thyme, and plenty of salt and pepper.

● You can place the crumbing in a plastic bag, add the food to be crumbed, and simply shake until coated. This works well in general, but be careful with fragile foods such as some flaky white fish fillets.

● British chef Gary Rhodes uses oatmeal for coating salmon. His glamorous presentation, ideal for dinner parties, involves dipping one side of the salmon in beaten egg and then coating only the eggy side with the oatmeal.

Kasha

Kasha means different things to different people. In shops today the word is often used to denote ready roasted buckwheat as opposed to raw buckwheat groats. Others use the term simply in place of buckwheat. According to Lesley Chamberlain's *The Food and Cooking of Eastern Europe*, however, kasha has traditionally referred to gruels and porridges, often made with buckwheat but also with millet or barley. Some people routinely toss the buckwheat in beaten egg before frying, which helps to keep the grains separate in the finished dish.

ingredients

½ tbsp vegetable oil, butter or duck fat

50g/¼ cup buckwheat groats

about 175ml/¾ cup chicken, fish,
   meat or vegetable stock

salt and pepper

per person

method

Heat the oil, butter or duck fat in a saucepan. Add the buckwheat and cook, stirring, for 5 minutes, or until the grains become orangey brown in colour and fragrant.

Pour in the stock and bring to the boil. Reduce the heat and simmer for 12–15 minutes or until the liquid is absorbed and the buckwheat is just tender.

Fluff up the grains with a fork and season to taste before further use or serving.

cook's notes

● To make a salad, add a handful of
flat parsley leaves, 10 torn mint leaves,
2 chopped spring onions, 5cm/2in seeded
and sliced cucumber, plus a squeeze of
lemon juice and salt and pepper to taste.

● While the buckwheat is cooking, sauté
2 thick rashers of bacon, finely chopped,
in a little oil or butter. Stir into the cooked
buckwheat with 2 finely chopped spring
onions and some parsley and chives.

● Make a stuffing for bream or other long,
round white fish by cooking double the
quantities given on the left. Stir in a finely
chopped onion fried in butter, a chopped
hard-boiled egg and some herbs such as
parsley, dill, oregano and chives, then heat
through. Stuff the fish, place in
a baking tin, pour some melted butter
over it and bake at 180°C/350°F/Gas 4
until the fish is done. Stir 2–3 tbsp of
soured cream into the cooking juices in
the tin to make a sauce.

● Dean and Deluca have devised a
fabulous potato gratin made using five
times the quantity of ingredients given left,
and chicken stock. While the buckwheat is
cooking, sauté 250g/9oz chopped
mushrooms and a sliced garlic clove for
10 minutes. Finely slice 1kg/2lb 4oz
potatoes and 250g/9oz onions. In a large
greased baking dish, place a layer of
potatoes, then scatter with the onion,
mushrooms and buckwheat, season as
you go. Cover with the remaining potato.
Pour 350ml/1½ cups double cream over,
then top with grated fontina or Gruyère.
Bake at 180°C/350°F/Gas 4 for 1 hour or
until the potatoes are tender. As they say, it
might not be traditionally Swiss, but it
should be!

Savoury millet cake

So proper chefs won't touch millet, eh? The idea for this savoury cake, from a book by internationally renowned chef Jean-Georges Vongerichten, is credited to his colleague Didier Virot. It's an extremely versatile side dish but also something I like to serve as a snack, covered with sauces or cream and piquant toppings. The eggy mixture reminds me of the sweet millet dumplings served at Kozue restaurant in Tokyo, and I think that with a little sweetening it could indeed be fried in walnut-sized balls and served with fruit sauce.

ingredients

200g/1 cup millet

450ml/2 cups water

4 tbsp butter, diced

3 eggs

4 tbsp milk

1 tbsp vegetable oil, plus extra for brushing

salt and pepper

serves 4

cook's notes

● The original recipe in Jean-Georges's book, co-authored with American food writer Mark Bittman, uses 2 whole eggs and 2 egg yolks. I'm not one for saving and reusing egg whites, hence the alteration, which works fine.

● They advise that while this dish 'can be served with almost anything' it is especially good with stews. My favourite uses are with roast chicken and a sauce of reduced chicken stock, or simply smeared with mascarpone or thickened crème fraîche and topped with flakes of smoked or roast salmon (creme fraîche and roast salmon pictured left) – even honey roast salmon bought ready-made at the supermarket.

● As long as you're careful, it is possible to leave the oven off, turn the half-cooked cake onto a plate and ease it back into the frying pan. This helps to give a crisp, moist finish.

method

Place the millet in a small saucepan with the water and a little salt. Bring to the boil. Skim the surface, then cover, reduce the heat right down and cook for 20 minutes or until all the water has been absorbed.

Remove the pan from the heat and leave the millet to stand, covered, for 10 minutes. Meanwhile, heat the oven to 240°C/475°F/Gas 9.

Stir the butter into the hot millet, so that it melts into the grain. In a small bowl, beat the eggs lightly and add them to the millet along with the milk, some salt, and plenty of freshly ground pepper. Mix to give a thick batter.

Place a small, ovenproof non-stick frying pan, or a flameproof dish, over a high heat on the stovetop. When it is very hot, add the oil. Then, when you can see the haze rising from the oil, spoon the batter into the pan. Smooth over the surface to shape it into a large, round cake.

Cook for 3 minutes or until the edges of the cake start to firm up and the underside is golden. Transfer the pan to the hot oven and bake for 5 minutes. Carefully brush the setting surface of the millet cake with oil, then continue cooking for 3–5 minutes or until it is firm on top but not dry.

Remove the pan from the oven and carefully invert the millet cake onto a large serving plate so that the crisp-fried golden side is uppermost.

Skirlie

The name 'skirl-in-the-pipes', as this moreish onion and oat mixture from Scotland is traditionally known, compares the sound of it sizzling in the frying pan to the whirring noise of bagpipes. The dish is a deconstructed white mealie pudding, made by frying the ingredients together rather than packing them into beef intestines to make sausages. I was most tempted to revisit it not through researching old recipes, but by a picture of skirlie-stuffed roast lamb in Sue Lawrence's mouthwatering book *Scots Cooking*, a stylish, modern take on the subject.

ingredients

2 tbsp butter

2 tbsp olive or vegetable oil

1 onion or leek, finely chopped

100g/1 cup medium or coarse oatmeal

salt and pepper

serves 4–6

method

In a large frying pan, melt the butter and oil together over a moderate heat. Add the onion or leek and reduce the heat right down. Cook, stirring often, for 10 minutes or until soft and golden.

Mix in the oatmeal and cook, stirring, until all the fat is absorbed. Add some more oatmeal if there seems to be too much oil in the pan.

Cook, stirring frequently, for another 8–10 minutes or until the mixture is well toasted. Season to taste with salt and pepper before serving.

cook's notes

● Traditionally this would be made with melted beef suet, or meat or bacon dripping, in place of the butter and oil.

● Try serving skirlie sprinkled over buttery mashed potatoes, combined with swede, if desired, or, as Sue suggests, over roast game birds in place of fried breadcrumbs.

● To use the mixture as a stuffing for roast leg of lamb, first have your butcher tunnel-bone the joint for you. Prepare the skirlie and leave it to cool, then pack it firmly into the hole created in the meat when the bone was removed. Tie along the joint with string to keep it together. Roast at 220°C/425°F/Gas 7 for 20 minutes before reducing the oven to 190°C/375°F/Gas 5 and cooking for another hour to give medium/well-done lamb. Always remember to leave roast joints to rest for 15 minutes before carving.

● Cut the tops off some beefsteak or other large tomatoes and use a spoon to scoop out the seeds and flesh. Fill the tomatoes with the skirlie, adding some of the chopped flesh if a particularly juicy result is desired, then bake in a moderate oven for 10–15 minutes until piping hot. Sprinkle the tops with parsley and, if you like, gently peel away the curled skin from the tomatoes before serving.

● The skirlie mixture can also be packed into a pudding basin or cloth and steamed to serve as a side dish.

Uppma

Here is an Indian-style treatment of millet that works well as part of a feast featuring several small dishes served together, especially one for vegetarians. The combination of whole spices, urad dal and curry leaves that flavours the dish at the start of cooking is typical of the Kerala region, though the recipe is not an authentic Indian one. I'm astonished that curry leaves are not commonplace in Britain, given the national passion for Indian food. The leaves freeze very well, however, so it's convenient to buy a large bunch from specialist grocers.

ingredients

3 tbsp vegetable oil or ghee

1 tsp mustard seeds

1 tsp cumin seeds

1 tsp urad dal lentils

1 small dried red chilli

a few curry leaves

2 green chillies, sliced

1 medium onion, finely chopped

2.5cm/1in cube root ginger, finely chopped

1 red bell pepper, chopped

75g/3oz broccoli, chopped

400g/2 cups millet

700ml/3 cups water

10g/ 1/2 cup packed coriander leaves

salt and pepper

serves 8

method

In a large pot, heat the oil or ghee, then stir in the mustard and cumin seeds, urad dal, dried red chilli and curry leaves. When the mustard seeds begin to pop, add the green chillies, onion and ginger and cook over a medium to low heat, stirring frequently, for 10 minutes or until the onion is well softened.

Add the bell pepper, broccoli, millet and some salt and pepper. Continue cooking and stirring for another few minutes until the millet is slightly toasted, then pour in the water. Bring to the boil, cover the pan and simmer gently for 20 minutes or until the water has been absorbed and the millet is puffed, fluffy and tender.

Stir in the coriander leaves and season the mixture to taste with salt and pepper. Fork through the millet and serve hot.

cook's notes

● Don't hesitate to use only the crunchy stalks of the broccoli in this dish, and save the florets for another recipe where their attractive appearance will make a difference.

● Swap other vegetables for the broccoli and bell pepper, as desired. A few handfuls of spinach stirred in towards the end of cooking would be good; don't put it in any earlier or it will become slimy.

● This recipe will work fine with long-grain white rice or quinoa, or with bulgur wheat.

● Dried beans or lentils, cooked separately and stirred in towards the end of cooking, would turn this into a complete, one-bowl meal.

● You could also serve it alongside simple lemony grilled white fish, the paler oily fishes such as marlin and kingklip, or chicken. Tandoori or tikka chicken and fish would also work well.

Pumpkin and sage pudding

There was an old *Spitting Image* series of sketches in which the grey puppet version of former British prime minister John Major would say 'Nice peas, Norma' to his wife at dinner every evening in a pathetic attempt to start conversation. In Australia we do a similar thing, observing after a long period of silence: 'That's a nice, dry bit of pumpkin.' Strongly flavoured dry pumpkin is what we like and is the best choice for this recipe. Butternut squash is just not as good. This recipe is based on one from Italian writer Emanuela Stucchi.

ingredients

100g/½ cup butter, plus extra for greasing

300g/11oz firm pumpkin flesh,
    peeled and deseeded

75g/⅓ cup rolled oats

2 tbsp finely chopped fresh sage

120g/¾ cup plain wheat-type flour

½ tsp baking powder

2 eggs, beaten

2 tbsp milk

salt and pepper

serves 4–6

method

Heat oven to 180°C/350°F/Gas 4. Grease an ovenproof baking dish. Using a food processor, grate the pumpkin into fine shreds.

In a large frying pan or saucepan, melt the butter and add the grated pumpkin, rolled oats and sage. Cook over a moderate heat for about 10 minutes, stirring frequently. Set aside to cool.

Combine the flour, baking powder and some salt and pepper in a large mixing bowl. Make a well in the centre and add the eggs and milk. Beat lightly, gradually incorporating the flour to make a batter.

Stir in the cooked pumpkin and oats, then transfer the mixture to the greased oven dish and bake for 30 minutes or until firm. Serve hot.

cook's notes

● Root vegetables, including carrot, swede and parsnip, can be used in place of pumpkin. Vary the herbs according to your personal tastes but choose robust varieties such as parsley, thyme, oregano, marjoram and rosemary.

● This pudding can be served as a side dish with chicken in various forms. A classy presentation is to use it as a base for roast chicken and a sauce of intense reduced stock or demi-glace.

● Vegetarians will enjoy it as part of a main meal, served, for example, with some cheese and a salad of broad beans and asparagus, or a sauté of peas, onion and lettuce.

● It also makes a nice savoury teatime treat or snack, and is a welcome addition to a cooked breakfast with eggs and fish or ham.

● Replacing the rolled oats with flakes of other grains is an option but not quite as successful because oats have a special creamy flavour and texture. Wheat flakes are the best second choice. Barley and rye are okay but very dry. You could, of course, combine any of the above for a mixed grain version.

Ogokbap

This recipe comes from my colleague Marc Millon, author, with wife Kim (who is a food photographer), of the intriguing *Flavours of Korea*. They spent much time travelling in this little-known country and feature many recipes from Marc's Korean grandmother, including several that mix rice with other grains. While this was a common means of stretching the rice supply, Marc says that this dish 'is by no means a poor person's rice, but a very special and splendid one.' Fans of Jamaican red beans and rice or 'Moors and Christians' will love it.

ingredients

50g/¼ cup dried black kidney beans

50g/¼ cup dried small red beans or aduki beans

200g/1 cup medium-grain white rice

100g/½ cup sticky rice

50g/¼ cup white pearled barley

50g/¼ cup millet

serves 8–10

method

Soak the black and red beans together in a bowl of water for 3–4 hours. Meanwhile, wash the two rices thoroughly. Place them in a separate bowl with the barley and millet, then cover them with water and soak for 2 hours.

In a pan of fresh boiling water, cook the beans for 30–45 minutes or until tender. Drain, reserving the cooking liquid, and set aside. Next drain all the grains, discarding their soaking liquid.

Put the beans and all the grains in a heavy pot. Measure the bean cooking liquid in a jug and make it up to 800ml/3½ cups with fresh water. Add this to the pot and bring to the boil. Give it a stir, then reduce the heat, cover and simmer for 30 minutes until the water is absorbed and the grains are tender.

cook's notes

● Korean grocers often sell bags of mixed grains featuring whole rather than refined rices, plus peas and pulses ready to cook. The pack I have in my kitchen includes an astonishing range: barley, millet, two types of brown rice, black sticky rice, Job's tears, red and black beans, mung beans, dried corn and dried peas. The problem is getting the pulses to cook as quickly as the rice. That's not an issue with this recipe because here the beans are cooked first.

● There are no salt or flavourings used in this dish, in the expectation that it will accompany strong-tasting, saucy main dishes that are very likely to be made with soy sauce. Add salt if you like, but not to the beans while they are cooking as this will toughen them.

● If your barley is semi pearled and beige rather than polished until white, you will need to preboil it for about 15 minutes before combining it with the rice and beans in the pot. If you don't, the rice will disintegrate before the barley is cooked.

Braised barley with lemon and spices

Preserved lemons and other bright-tasting aromatics here lift what are normally considered to be two very humble ingredients – barley and celery – out of the ordinary. Europeans rarely cook with their cultivated variety of celery, regarding it as a salad vegetable. However, braising is an excellent treatment for it and easy, to boot. When you lift the lid off the oven dish and give the mixture a brief stir before serving, in addition to smelling the wonderful fragrance, you will see it has become luxuriously creamy and comforting, quite like a risotto.

ingredients

275ml/1¼ cups vegetable, chicken or lamb stock

50g/¼ cup semi-pearled or pot barley

6 stalks celery, chopped

1 piece preserved lemon rind, chopped

2 tbsp lemon juice

1 tbsp olive oil

4–6 sprigs thyme, or a bunch of parsley
    and coriander stalks

1 bay leaf

10 peppercorns, crushed

10 coriander seeds, crushed

1 clove garlic, crushed

a pinch of chilli flakes

serves 2–4

method

Heat the oven to 180°C/350°F/Gas 4. Put all the ingredients in a casserole. Cover and bake in the hot oven for 1½ hours or until the barley is tender and most of the liquid has been absorbed.

Remove the casserole dish from the oven and stir the mixture briefly to combine the ingredients. Pick out the large flavourings before serving.

cook's notes

● This dish is a good accompaniment to roast or grilled chicken and lamb. Vegetarians will be happy to eat it as a main course.

● For a simple British barley braise to serve with roast meat, combine the barley, celery, bay and thyme in a casserole with 300ml/1⅓ cups lamb stock and cook as instructed here.

● Also suitable for this recipe are black barley, plump wholegrain rice, freekeh and semi-pearled wheatberries such as farro. Polished short or medium-grain rice would be good, too, but cut the cooking time and quantity of stock a little.

● This dish can be quite fiery, depending on the size of your 'pinch' of chilli flakes. If your tolerance to heat is low, a few flakes will do.

● Preserved lemon is a lovely inclusion but feel free to skip it if you have none to hand. Australian Maggie Beer's brand of preserved lemons are now as easy to buy in London as they are in the mother country. They are excellent and helpfully contain a plastic disc that keeps the lemons tucked down under the pickling juice. When using preserved lemons, trim away the flesh of the fruit and use the rind only. People cut up lemons for preserving in different ways. Here '1 piece' of preserved lemon rind, means about a quarter of what would have been a whole lemon.

● The stock and preserved lemons make this dish salty enough; there's no need to add more.

African corn and peanut patties

Africans use a wide variety of dark greens in the manner of spinach. Morogo and 'African spinach' are terms that cover a group of leaves that may include those cut from pumpkin and sweet potato plants, as well as silverbeet and edible wild greens such as buffalo-thorn. Peanuts, or groundnuts, are technically not nuts at all but the seeds of a legume similar to soy. They are characteristic of West African cuisine, having been taken there from their native South America by the Portuguese. Ironically, it was the Africans who took them to the USA.

ingredients

50g/ ¼ cup samp or hominy

75g/ ⅓ cup coarse cornmeal

150g/5oz spinach leaves, shredded

¼ tsp ground turmeric

¼ tsp salt

a sprinkle of chilli flakes

275ml/1 ¼ cups boiling water

25g/ ¼ cup chopped roasted and salted peanuts

pepper

a little butter and olive or vegetable oil, for frying

serves 2–4

cook's notes

● Instead of shaping the mixture into patties, you could leave it soft and serve it as a textured alternative to cornmeal dishes such as polenta.

● The corn and nuts complement each other to form a complete protein dish, so this recipe can be used as the base of a vegetarian meal.

● Serve with a bright-tasting, chilli-laden tomato sauce or relish, with or without some barbecued meats and poultry. The tomato sauce layered with cornmeal to make Pap Tart (page 124) would be a lovely accompaniment.

method

Soak the samp or hominy in a bowl of cold water overnight or for at least 4 hours. Drain and rinse.

In a large, heavy saucepan, combine the soaked corn, the cornmeal, spinach, turmeric, salt and chilli flakes. Stir in the boiling water and place the saucepan over a low heat. Cook, stirring frequently, for 20–30 minutes, adding more hot water to the pan as necessary.

When the mixture is very thick, remove it from the heat. Stir in the peanuts and pepper to taste. Set aside until the mixture is cool enough to handle. Shape it into patties using floured hands, then fry in a little butter and oil until browned on both sides.

Sicilian-style farro and ricotta pudding

Sicilian culinary experts will exhort that this is not *cuccìa*, their traditional whole wheat and ricotta pudding, and they are right. It's an updated version of that intensely flavoured, heavy and yet quite sticky dessert. The whole wheatgrain is swapped for a reduced quantity of semi-pearled farro, which is much quicker and easier to prepare as well as being lighter on the palate. The quantity of sugar has also been reduced considerably to suit contemporary tastes. It's still an acquired taste, but after half a bowl you may well decide you love it.

ingredients

75g/¾ cup semi-pearled farro or other
    wheat-type grain

250g/9oz ricotta

1½ tbsp caster sugar

1½ tbsp amaretto or 1 tsp vanilla extract

2 tbsp chopped candied citron or candied pumpkin

20g/¾oz fine dark chocolate, roughly chopped

serves 4

method

In a small saucepan, cook the farro in a generous quantity of boiling water until just tender – usually around 25–30 minutes. Rinse under cold running water and set aside to drain thoroughly.

In a mixing bowl, whisk the ricotta and sugar until smooth, light and creamy, then briefly whisk in the amaretto or vanilla extract. Fold in the candied citron or pumpkin and the chopped chocolate.

Spoon the mixture into serving dishes or ramekins and leave to stand for an hour at room temperature before serving. Alternatively, chill the puddings then turn them out and garnish with candied orange peel and sifted icing sugar if desired.

cook's notes

● Even with the modifications made to the traditional pudding, there are many people who will prefer it refined further. Omitting the candied fruit and chocolate altogether gives a lovely textured cream to serve with fresh fruit such as figs or sliced oranges.

● Candied citron or pumpkin can be bought from specialist Italian stores or good food halls. The candied peel sold in supermarkets for fruit cakes and other home baking is not preferred here. Better is to substitute other glacé fruit if necessary.

● Many Italians would make this with cooked canned wheat rather than soaking and cooking the wholegrains from scratch.

● Mary Taylor Simeti, author of several books on the subject, is adamant in *Sicilian Food* that the ricotta should never be beaten with an electric whisk. I am not against it.

● Traditional Neapolitan wheat tart features a filling reminiscent of *cuccìa*. To make it, line a tart case with rich shortcrust pastry, saving some excess pastry to cut into strips for a lattice top. Soak 200g/1 cup whole wheat for 24 hours, then gently cook it for 3–4 hours in 570ml/2½ cups of milk flavoured with pared lemon zest. When done, flavour with 2 tsp caster sugar, 1 tsp ground cinnamon, a few drops of vanilla extract and the zest of half an orange. Beat together 300g/10oz ricotta, 4 egg yolks and 4–6 tbsp orange flower water. Fold in 150g/5oz candied peel and the grain mixture. Make a meringue from the 4 egg whites and 200g/1 cup caster sugar and fold them in too. Fill the tart case, cover with the pastry strips in lattice formation, and bake at 200°C/400°F/Gas 6 for 45 minutes. Serve cold, dusted with powdered sugar.

Milhassou

Although it is most commonly made with cornmeal these days, milhassou is a type of old Spanish and French country pudding traditionally made (the name is a clue) from millet. This grape-studded custard-and-clafouti-style version tastes strongly of honey and lemon, and as it is rich with eggs and milk, smoothes over any bitter taste that may be present in the millet flour. It makes a good Sunday lunch dessert, or when cooked in ramekins, a comforting one to eat on a Sunday evening in front of the television.

ingredients

150g/1 cup grapes, de-pipped if necessary

2 tbsp rum, brandy or eau de vie

600ml/2½ cups milk

4 eggs

75g/⅓ cup millet flour

250g/¾ cup runny honey

finely grated zest of 1 lemon

butter, for greasing

serves 6

cook's notes

● Replace the grapes with fruit such as cherries, berries or sliced plums and nectarines if desired. Alternatively, you can omit the fruit and alcohol altogether if you prefer, and use orange zest in place of lemon.

● Instead of dividing the mixture among individual ramekins, cook it as one large pudding for 20 minutes at the same temperature.

● As ever, the fresher the millet flour the better the result. You can also make this pudding using coarsely ground cornmeal, or with wheat semolina.

method

Place the grapes in a mixing bowl and cover with the alcohol. Set aside to marinate for about 1 hour.

Heat the oven to 220°C/425°F/Gas 7. Grease 6 small ovenproof ramekins or soufflé dishes of about 250ml/1 cup capacity and divide the marinated grapes between them.

In a heavy saucepan, slowly bring the milk to scalding point. Meanwhile, in a mixing bowl, whisk together the eggs, millet, honey and lemon zest. When the milk is hot, slowly pour it over the egg mixture, whisking constantly.

Pour the custard mixture into the ramekins. Place them on a baking tray and cook for 10–12 minutes until only just set and golden. Eat warm or cold.

Sticky black rice custard

Sticky black rice must be the most fashionable wholegrain, featured as it often is on the menus of trendy fusion restaurants with award-winning chefs as well as contemporary pan-Pacific noodle houses with their communal tables and paper place-mats. Yet many who eat it don't realise the long, elegant, purply-black grains are whole and doing their insides some good. It has a lovely flavour and texture but unfortunately can lend a pinky-grey hue to ingredients it is cooked with, as is the case here. Simply close your eyes and think of Thailand.

ingredients

3 tbsp butter

100g/½ cup sticky black rice

340ml/1 ½ cups coconut milk

125ml/½ cup milk

100g/½ cup caster sugar

1 vanilla pod, split

3 egg yolks

4–6 tbsp desiccated coconut

serves 4—6

cook's notes

● This dish can be made with hominy corn in place of the sticky black rice.

● It's rich but moreish – rather too so. Serve it plainly and in small portions. If you think you must add something else, choose simple fresh fruit from the tropics such as mango and use the black rice (or corn) custard almost like a scoop of ice cream served on the side.

method

Melt 1 tbsp butter in a large, heavy saucepan over a moderate heat. Add the rice and stir to coat. Add the coconut milk, milk and sugar and stir to combine. Scrape the seeds from the inside of the vanilla pod and add them to the pan along with the pod. Bring the mixture to the boil, then reduce the heat and cook gently for 10 minutes.

In a large mixing bowl, whisk the egg yolks. Add a ladle of the hot milk mixture to the yolks and whisk to combine. Pour the yolk mixture into the saucepan of milk mixture, stirring constantly, and cook for 2 minutes. Stir in the remaining butter. The mixture will be very thick.

Pour into a serving dish and set aside to cool. Cover the cooled custard with clingfilm to prevent a skin forming and chill for 2 hours. Meanwhile, in a heavy frying pan, toast the desiccated coconut over a medium to low heat for 2–3 minutes, stirring constantly until the coconut is fragrant and has an even golden brown colour. Immediately tip the toasted coconut into a small bowl and set aside to cool.

When ready to serve, cover the black rice custard with an even layer of the toasted coconut.

Wheatberry and honey cake

If you like solid, comforting fruit and vegetable based cakes, you will probably love this one. The unpolished wheatberries add a hearty texture but some also seem to disappear mysteriously into the cake batter, so that the texture is not as sturdy as a cupful of grain may suggest. This recipe is based on one found in South Africa, where semi-pearled wheatberries are readily available in supermarkets as stampkoring. The flavour is a little reminiscent of Britain's treacle tart, with only a hint of ginger. If you love this spice, use more.

ingredients

50g/¼ cup semi-pearled wheatberries

110g/½ cup butter, plus extra for greasing

90g/½ cup brown sugar

1 egg, beaten

300g/2 cups plain wheat flour

2 tsp baking powder

½ tsp ground ginger

125ml/½ cup honey

125ml/½ cup milk

½ tsp salt

serves 12

cook's notes

● Wild rice can be used in place of the wheatberries, if you prefer, or just replace a portion of the wheat with wild rice. Any grains that find their way into the crust of the cake will become hard and crunchy.

● Mixing in 2 medium overripe bananas, about 240g/1 cup mashed, gives a lovely banana cake that is delicious with wild rice. If you want to retain a spicy flavour, you need to bump up the ginger and/or add some more spices to compensate for the strong taste of the fruit.

● You can use golden syrup or treacle instead of the honey if preferred.

method

Place the wheatberries in a small saucepan. Cover generously with water and bring to the boil. Reduce the heat and simmer until tender – the exact time will depend on the variety and degree of pearling. When cooked, drain and rinse the wheatberries, then measure them: you should have about 125ml/1 cup. Set aside to cool.

Meanwhile, grease a loaf tin with butter and line it with baking paper or foil, leaving some excess hanging on each side to help you lift the cake out later. Then grease the lining paper or foil as well.

Heat the oven to 180°C/350°F/Gas 4. In a large mixing bowl, beat the butter and sugar together until creamy and light. Gradually beat in the egg, then about half the flour, plus the baking powder and ginger. Beat in the honey, then gradually add the milk and the remaining flour, alternating between them.

Fold in the cooked wheatberries and transfer the cake batter to the prepared tin. Smooth over the surface, then make a furrow down the centre of the batter to encourage even rising. Bake for 60–75 minutes or until a skewer inserted in the centre of the cake comes out clean.

Remove the cake from the oven and leave to stand in the tin for at least 10 minutes before lifting it out to cool further on a wire rack. Serve warm or cold.

Torta Sbrisolona

The first torta sbrisolona I ever ate was produced by Italian baker Giuliana Lamburghini, who is based in Rome but comes from Mantua, where this cake originates. It was delightful and, rightly or wrongly, I've compared all subsequent versions I've sampled to that most enjoyable experience. Some people prefer a softer, cakier style than the recipe featured here, which is like a large cornmeal and almond shortbread. It's traditionally smashed in the centre to be devoured in crumbling chunks with sticky wines or grappa, much as cantuccini are eaten.

ingredients

300g/2 cups plain wheat flour

100g/²⁄₃ cup coarse yellow cornmeal

100g/²⁄₃ cup fine yellow cornmeal

150g/1 cup almonds, chopped

200g/1 cup caster sugar

finely grated zest of 1 lemon

225g/1 cup butter, cubed, plus extra for greasing

3 eggs

1 tbsp vanilla extract

sifted icing or powdered sugar, for dusting

serves 4–6

method

Heat the oven to 180°C/350°F/Gas 4 and grease a 25cm/10in round or square cake tin. In a large mixing bowl, combine the plain wheat flour, the two cornmeals, plus the almonds, sugar and lemon zest. Use your fingertips to rub the butter into the dry ingredients until the mixture resembles breadcrumbs.

In a small bowl, beat the eggs and vanilla extract together. Make a well in the centre of the flour mixture and pour in the eggs. Give it a brief stir, then quickly finish working the dry ingredients into the wet using your hands to give a solid, pastry-like dough. Do not knead the dough; work it until it only just comes together.

Press the dough into the prepared tin. Smooth the surface over a little but leave it dimpled as this will enhance the crumbly texture. Bake for 35–40 minutes until the top is just starting to brown. Leave the cake to cool in the tin. Dust with sifted icing sugar and break into chunks to serve.

cook's notes

● Hazelnuts are sometimes used in place of, or in conjunction with, the almonds.

● For extra shortness, replace half the butter with the same quantity of lard.

● Some Italian cooks believe that only finely ground cornmeal should be used in this cake.

● It's important (I think) that the cake is not too thick – about 2.5–3cm/1–1½in deep in the tin is best. If it's any thicker, the result will be too scone-like. If you have any excess, shape it into walnut-sized balls and bake as cookies, which far from being a second-rate use of leftovers have become my partner's favourite. You could, of course, shape the whole mixture into cookies. They will need only 20 minutes or so in the oven.

● The quantity of vanilla may seem excessive but do not skimp here – it is important to the flavour. Also, be sure never to use vanilla essence or 'flavouring' in cooking. Though much cheaper, it is a substantially inferior product to pure vanilla extract, for which there are regulations governing production. The true intoxicating aroma and rich taste of vanilla are, as Alan Davidson's *Oxford Companion to Food* states, 'unreproducible in the laboratory'.

Crunchy oat and coconut slice

In Britain these would be called flapjacks, a term very confusing to the rest of the English-speaking world because they quite clearly don't flap. At my first job in London I drove the man with the sandwich trolley crazy for 18 months because I was unable to grasp the concept of a thick, stiff, resolutely unflappable flapjack. Seeing that they were made from oats, I made matters worse by calling them oatcakes. This melt-and-mix recipe is terrific for children, which I say with some authority because it was one of the first things I ever made.

ingredients

225g/1 cup butter
250g/2 cups rolled oats
160g/1 cup brown sugar
100g/1 cup desiccated coconut
serves 12

method

Heat the oven to 170°C/325°F/Gas 3. In a large saucepan, slowly melt the butter over a gentle heat. Using a wooden spoon, stir in the oats, sugar and coconut, mixing until well combined.

Transfer the mixture to a 20cm/8in square tin and press it out evenly. Bake for 20–25 minutes or until light golden. Leave to cool in the tin, then cut into pieces and serve.

cook's notes

● To make fruited versions of this slice, simply stir in about 75g/ ½ cup of sultanas or raisins, chopped dried apricots, prunes, mango or other dried fruit at the same time as adding the oats, sugar and coconut.

● Walnuts are a terrific addition, but other nuts will also work well. Add about 60g/ ½ cup to this quantity of mixture.

● Warm spices such as cinnamon, nutmeg and ginger can also be added. Use 1–1 ½ tsp in total of your favourites and add them along with the dry ingredients.

● Don't use shredded coconut or coconut flakes. It might seem as though they will look more attractive, but they make the mixture less absorbent than desiccated coconut, so the baked result is greasy and uncohesive.

● If the corners of the slice are browning too quickly, cover them with some foil.

Bellies

Uncooked amaranth adds a wonderful crunchy texture to these traditional Mexican cornmeal cookies, which really appeal to children and adults, whether health food fans or not. Sunshine-yellow masa harina, in which the grain has been alkali treated, is a beautifully fine and light flour that can be used in all sorts of baking and gives a vibrant colour to the finished dish. Why these are called bellies I don't know. You'll certainly get one if you eat too many, but it might be a reference to the dried fruit, which could be said to resemble a navel.

ingredients

150g/⅔ cup butter, plus extra for greasing

160g/¾ cup caster sugar

3 eggs, beaten, plus 1 egg yolk, beaten, to glaze

300g/2 cups masa harina

1 tsp baking powder

260g/1 cup amaranth

4 tbsp raisins or sultanas

makes 24

method

Heat the oven to 180°C/350°F/Gas 4 and lightly grease a large baking sheet.

In a large mixing bowl, beat the butter and sugar together until light and fluffy. Beat in the whole eggs then quickly start to beat in the masa harina and baking powder. Once a dough has been formed, add the amaranth and mix well.

Grease you hands with a little butter. Shape the mixture into balls about the size of a walnut and press a raisin or sultana into the centre of each. Set them well apart on the baking sheet and glaze with the beaten egg yolk. Place in the oven to bake for 20 minutes or until golden brown.

Remove from the oven and transfer to a wire rack to cool. Store in an airtight container.

cook's notes

● It is possible to use regular yellow or white cornmeal in this recipe, but choose a fine grind as there is already enough grittiness in the mixture thanks to the amaranth.

● Cornmeal cookies are also traditional in countries such as France and Italy. For Italian biscotti to serve with coffee, rub together 150g/1 cup stoneground cornmeal, 50g/⅓ cup plain wheat flour, 110g/½ cup butter and 1 tsp lemon or orange zest until the mixture resembles breadcrumbs. Beat 1 large egg with 100g/½ cup sugar and 1 tsp vanilla extract and stir this mixture into the dry ingredients to give a sticky dough. Shape into balls, space them well apart on a greased baking sheet and bake at 190°C/375°F/Gas 5 for 15 minutes or until the edges of the cookies start to brown.

Anzac biscuits

During the first World War, it quickly became common usage to refer to the soldiers of the Australian and New Zealand Army Corps by its acronym. The first Anzac Day, which remains an annual remembrance event Down Under, was inaugurated on 25 April 1916 to commemorate the first anniversary of the landing of the Anzac troops at Gallipoli. These biscuits, devised by the women at home to send to the fighting men, had to be able to survive at least two months at sea on merchant navy ships, so are chewy and made without eggs.

ingredients

85g/1 cup rolled oats

150g/1 cup plain wheat flour

100g/ 1/2 cup sugar

50g/ 1/2 cup desiccated coconut

110g/ 1/2 cup butter, plus extra for greasing

1 tbsp golden syrup

1/2 tsp bicarbonate of soda

1 tbsp boiling water

makes 30

method

Heat the oven to 150°C/300°F/Gas 2 and lightly grease a baking sheet. Place the oats, flour, sugar and coconut in a mixing bowl.

In a small saucepan, melt the butter and golden syrup together over a low heat. Meanwhile, place the bicarbonate of soda in a small bowl and pour on the boiling water. When the butter has melted, stir the soda solution into the saucepan so that the mixture foams up.

Pour the bubbly butter mixture into the bowl of dry ingredients and mix well to give a chunky dough. Place tablespoons of the mixture on the greased baking sheet, spacing them well apart.

Bake the Anzacs for 20 minutes, then remove them from the oven. Using a palette knife or fish slice, transfer the biscuits while still warm to a wire rack to cool before eating. Store in an airtight tin.

cook's notes

● Myriad recipes now exist for Anzacs, and they are sold mass-produced in Australian and New Zealand supermarkets with various flavourings, including macadamia nuts and wattleseeds. As with all traditional regional foods, every cook has a favourite recipe and tut-tuts at any version deemed to be inauthentic. I purse my lips and frown disapprovingly at Anzacs with a cakey texture. They are supposed to be chewy but softish.

● There is some dispute about whether coconut was an original ingredient – apparently it was not easy to buy in Australia during the First World War, or so some people claim – but it quickly came to be seen as one of the Anzacs' characteristic ingredients. The other essentials are rolled oats, golden syrup (the binding agent), butter and bicarbonate of soda.

Rye and cheese crackers

Cheese-flavoured crackers are a delight, though many people don't realize that they are best not served with fine cheeses – their flavour detracts too much from those that the artisan maker will have tried so hard to put into the cheese itself. But for nibbles with drinks and general snacking, cheese crackers are most welcome. This recipe is based on one from my friend Lorna Wing's first book on party food. The rye adds an East and Central European quality that makes this a good match for rich creamy pâtés, hams and mild cured sausages.

ingredients

125g/scant 1 cup rye flour, plus extra
  for dusting
95g/7 tbsp butter, plus extra for greasing
110g/1 cup grated mature cheddar
½ tsp cayenne pepper
2 tbsp poppy seeds
salt
makes 20

method

In a large mixing bowl, rub the flour and butter together until they resemble fine breadcrumbs. Mix in the cheese, cayenne pepper and poppy seeds, and knead lightly together in the bowl to form a dough. Wrap in plastic and chill for at least 30 minutes.

Grease a large baking sheet lightly and set aside. Dust a work surface with flour and roll out the chilled dough until about 4mm/¼in thick. Use a cutter to shape the dough into rounds, then use a palette knife or fish slice to transfer them to the baking sheet. Re-roll the trimmings, cut them into rounds and add to the baking sheet. Cover and place in the refrigerator for 30 minutes.

Heat the oven to 190°C/375°F/Gas 5. Place the crackers in the oven and bake for 8–10 minutes or until their naturally greyish colour is beginning to turn a golden brown.

Remove from the oven and transfer to a wire rack to cool. Store in an airtight container.

cook's notes

● Replace the poppy seeds with sesame seeds, or with some very finely chopped walnuts if you prefer.
● Any stoneground wholemeal wheat-type flour can be used in place of the rye flour. You could also use plain wheat flour, or the same quantity of beremeal or other barley flour.
● Mustard powder or ground black pepper can be used instead of cayenne, and you may wish to increase the quantity of both to give a spicier biscuit. A small amount of celery seed is another possibility here.
● Cheddar is not essential. Use any hard and strongly flavoured cheese in its place, such as gruyère, comté or pecorino.

Oatcakes

As Nigella Lawson says in *How to be a Domestic Goddess*, there is something very satisfying about making 'good plain fare such as oatcakes – as if you're doing something sober and basic and not entertaining yourself with fripperies'. Yet in many households where oatcakes are not the daily bread, as they once were in Scotland and Ireland, they have found their métier on the cheeseboard, the serving of which often indicates a proper dinner party. How chic it is to serve home-made crackers rather than shop-bought ones on these occasions.

ingredients

50g/ ¼ cup coarse oatmeal

50g/ ¼ cup medium oatmeal

50g/ ¼ cup beremeal or barley flour,
  plus extra for dusting

½ tsp salt

2 tbsp butter, plus extra for greasing

a little boiling water to mix

makes 12

method

Heat the oven to 150°C/300°F/Gas 2. Put some water on to boil in the kettle and lightly grease a baking sheet.

In a large mixing bowl, combine the oatmeals, beremeal and salt. Using a round-bladed knife, cut the butter into the mixture and stir to help distribute it. Make a well in the centre and pour in 1 tbsp or so of boiling water. Mix to give a stiff dough, adding more hot water if necessary.

Dust a work surface with the beremeal and roll out the dough into a thin square. Leave to rest for a few minutes, then cut the dough into squares or whatever your desired shape, trimming away any very craggy edges but leaving the rims of the oatcakes rugged as the grains break away.

Using a palette knife or fish slice, transfer the oatcakes to the baking sheet and bake for about 30 minutes until only just beginning to brown. Remove from the oven and leave to cool. Store in an airtight tin.

cook's notes

● This recipe is based on one by Catherine Brown. The chunks of oats and beremeal give a pleasingly rugged texture but you can make a smoother oatcake if you prefer, by using only medium or fine oatmeal.

● The beremeal can be replaced with the same quantity of fine oatmeal, or with wholemeal wheat or rye flour if desired.

● Melted dripping or bacon fat would often be used in place of butter. A mixture of lard and butter could also be used.

● Cut the oatcakes into rounds if you prefer, but the square shape offers an element of mild surprise that is welcome at dinner parties. It also stops you pfaffing around with dough trimmings.

● Oatcakes can also be baked on a griddle, girdle or heavy frying pan. They need to cook slowly and steadily, for a good 5 minutes, so that they dry out thoroughly. The pan needs to be solidly hot, but not fiercely so.

Popcorn

From the foreigner's perspective, it can seem as though there is nothing that Americans won't do with popcorn, or jello for that matter. Some recipes take a little getting used to (a popcorn version of bread and butter pudding springs to mind), especially as popcorn's texture is so naturally dry and crackly. Still, who of us doesn't associate a box of popcorn with a trip to the cinema? Home-made and flavoured popcorn can be a terrific indulgence and tremendous fun. Whether or not you make it low-fat and healthy is up to you.

Basic popcorn

ingredients
2 tbsp vegetable oil
about 50g/ ¹⁄₄ cup popcorn
serves 2–3

method
Place the oil in a large, heavy pot over a moderate heat. Pour in enough popcorn to cover the base of the pot in a single layer, then cover and leave to cook so that the popcorn pops, shaking the pot occasionally.

When the popping noises become infrequent, remove the pot from the heat and serve, adding flavourings as desired. Alternatively, use an air-popping machine according to the manufacturer's instructions.

cook's notes

● My favourite is to melt 2 tbsp butter
with 2 tbsp crunchy peanut butter and stir
it into the popcorn.

● Alternatively, mix with 4 tbsp grated
parmesan, 1 tsp chopped fresh oregano,
¼ tsp dried red chilli flakes and loads of
salt and pepper.

● You could pop the corn in chilli oil,
then sprinkle with 2 tbsp salt, 1 tbsp
ground cumin, plus dried red chilli flakes
and freshly ground pepper.

● For spicy fruit and nut popcorn, heat
4 tbsp oil in a wok. Add 175g/1 cup
medjool dates and 85g/½ cup dried
apricots, both roughly chopped, plus
250g/1¾ cup mixed nuts such as almonds,
cashews and macadamias. Stir-fry until
lightly browned, then mix in the popped
corn and turn until the ingredients stick
together a little. Turn off the heat. Stir in
the grated zest of 1 lemon or lime, a finely
chopped chilli, 2 tbsp coriander leaves,
1 tbsp torn basil plus salt and pepper.

● To make popcorn balls, melt 3 tbsp
butter in a large saucepan. Add 200g/7oz
chopped marshmallows and 3 tbsp brown
sugar and stir until melted. Remove from
the heat and mix in the popped corn and
125g/¾ cup M&Ms or Smarties. Butter
your hands, shape the mixture into balls
and place on parchment paper to dry.

● Make popcorn and peanut brittle by
stirring the popped corn and 150g/1 cup
salted peanuts into hot toffee sauce (page
201). Pack into an oiled and lined cake tin
(oil the lining paper too) and bake for
20 minutes at 180°C/350°F/Gas 4. When
done, spoon the mixture into a cold tin,
also oiled and lined. Leave until set, then
cut into pieces and serve.

Alegria

Real hippy food this. It is the supposedly virtuous type of 'no added sugar' snack that is a mainstay of natural food stores throughout the Western world, and yet is well rooted in the traditional cooking of Central and South America. 'Alegria' in Spanish means happiness, and many travellers to Mexico will have encountered this popped amaranth sweet, although this version is based on a recipe from the Palenqueno people of Colombia. It's the proper thing to eat while reading *The Celestine Prophecy*, levitating or meditating on your aura.

ingredients

25g/1 cup popped amaranth (see cook's notes)

4 tbsp desiccated coconut

1–1½ tsp aniseed

7-8 tbsp runny honey

makes 16

method

Place the amaranth, coconut and aniseed in a large mixing bowl and stir to combine evenly. Gradually and gently, stir in the honey to give a stiff but still bubbly dough without crushing the grains.

Have ready a double thickness of greaseproof paper. The mixture will be sticky. Using two dessert spoons, shape the grain mixture into balls about 2cm/¾in in diameter and place them on the paper, keeping them well spaced.

Leave to dry for several hours. During this time they will firm up on top, although little pools of honey will form at the base. Periodically, roll the balls over half a turn so they are not wallowing in the syrup. Scoop up any clumps of mixture that stick to the paper and simply press the balls back into shape with your fingertips and put them back to continue drying. The sweets will not become hard; they are ready when firm enough to pick up and eat using the fingers.

cook's notes

● Popping amaranth at home is easy to do but not easy to do well. You need to pop the grain in batches of only 1 tbsp at a time, making any substantial quantity tiresome to produce. Furthermore, many of the grains will burn before they pop and separating them from the others is a pain. Do yourself a favour and buy it industrially blasted with hot air by a proper factory.

● If you prefer to ignore this advice, use a frying pan or sauté pan (the amaranth doesn't jump very high) and have a bowl ready by the stove to hold the popped grains. Get the pan solidly hot, then add 1 tbsp of amaranth and stir constantly with a thick, natural bristled pastry brush. When most of the amaranth has popped, quickly tip it into the bowl, then add another tbsp of grain and repeat. You will have to do this 8 times to pop just ½ cup of grain, and some burning is inevitable.

● Other grains, including wild rice and millet, can be popped. Millet pops best in oil rather than a dry pan, but as it is such a small grain the problem is how to fish it out of the hot oil once popped – you need a very fine-meshed, long-handled scoop of some kind. Also key is to ensure that the grain is covered by the oil, unlike popcorn, which just sits on top of it. If there is not enough oil to cover the seeds, they will burn and the effect on the pan will be scary.

● Wild rice doesn't so much pop as poufs, swelling up like a crusty pillow. It tastes very nice, especially salted but, again, be sure that the grains are covered by the oil and, before deciding to make it, consider how you are going to scoop the finished grains from the oil.

Mealie candy

Home candy making is something of a lost art, but fun to do occasionally, especially as it allows you to use tasty, wholesome flavourings such as toasted oatmeal that most sweets manufacturers wouldn't touch with a bargepole. This recipe is much easier than old-fashioned barley sugar, which requires taking the sugar syrup to a higher temperature and working the hard mixture on a cool slab with a special spatula. The small investment in a candy or sugar thermometer will make this sweet seem even easier to make.

ingredients

50g/¼ cup coarse oatmeal

a little vegetable oil or butter, for greasing

500g/2½ cups caster sugar

300ml/1⅓ cups water

4 tbsp treacle

1 tsp lemon juice

2 tbsp finely chopped glacé ginger

2 tbsp finely chopped walnuts

makes 25

method

In a heavy saucepan, toast the oatmeal over a moderate heat, stirring constantly until the grains are browned and fragrant. Transfer to a bowl to cool and wipe out the pan.

Lightly grease a small cake tin. Line it with foil, leaving some excess hanging at the sides of the tin so that the candy is easy to remove later. Grease the foil lining and set the tin aside ready to pour the hot candy into it.

Combine the sugar, water, treacle and lemon juice in the saucepan. Place over a low heat until the sugar has dissolved, stirring occasionally, then bring the mixture to the boil. Boil vigorously without stirring for several minutes until the syrup reaches 115°C/240°F on a candy thermometer.

Immediately remove the pan from the heat and, with a wooden spoon or paddle, beat the syrup until it just starts to form a mass of little white specks but before it becomes very thick.

Working quickly, stir in the oatmeal, ginger and walnuts and pour the mixture into the prepared tin. Smooth over the surface and leave to cool. When the mixture has almost set, mark it into squares with a knife. To serve, lift the block of candy from the tin and break it up into pieces following the line of the cuts made earlier.

cook's notes

● The lemon juice adds some flavour but also plays an important role in preventing the early granulation of the sugar syrup. In its place you could use a pinch of cream of tartar.

● The specific temperature the sugar syrup needs to reach is also known as the 'soft ball stage'. It is possible to test for this without a candy thermometer but it does make working swiftly enough more problematic. To test for soft ball stage, drop a little of the syrup into a glass of cold water and leave to cool for a few minutes. The syrup is at the right temperature when the syrup can be picked out of the water and rolled into a soft ball between the fingers.

● Ginger and walnut are just two of the flavours that can be used in this type of candy. You may prefer other nuts or glacé fruit, or to add some coconut and orange zest. Maple syrup would also be a nice addition here.

● Preserved stem or crystallized ginger can be used in place of glacé ginger if necessary.

Oat warmer

The truth is: this is gruel. Not the horrible Oliver Twist kind of stuff, but a soothing, creamy, thickened mixture in the manner of the cosy bedtime drinks people tend to buy in instant powdered form and which so often taste chemical rather than comforting. In old cookbooks recipes for these types of drink are legion and usually found under a chapter on food for the sick. It's best, I find, to ensure that whoever you make this for has first put on their pyjamas and tucked themselves up under the duvet with Winnie and Tigger.

ingredients

2 tbsp fine oatmeal

about 340ml/1½ cups milk

a pinch of salt

1 tsp caster or granulated sugar

a little butter

a little grated nutmeg

serves 2–3

method

In a medium-sized heatproof jug, blend the oatmeal with 2 tbsp of the milk. Place the remaining milk and the salt in a lipped saucepan. Bring to scalding point, then pour the very hot milk slowly onto the oatmeal mixture, stirring constantly. Leave to stand for 1–2 minutes.

Strain the milk back into the pan, discarding any bits of oatmeal in the sieve. Bring the mixture to the boil, stirring constantly, then reduce the heat and simmer very gently for 8–10 minutes until thickened, stirring often to prevent scorching.

Remove the pan from the heat and stir in the sugar. If the mixture has become extremely thick, add some more milk to thin it to the desired consistency. Pour the hot drink into serving mugs, then float a tiny dab of butter in each and top with grated nutmeg before serving.

cook's notes

● For a chocolate version, simply stir a large square of dark chocolate into the hot drink before pouring it into cups. Skip the butter, but add a little cinnamon if desired.

● This recipe can also be made with barley meal, cornmeal or white rice. Add your own favourite sweet warm spice to flavour.

● A traditional British country drink called caudle replaces the milk in this recipe with equal quantities of light beer and water, and adds ginger and cloves to the nutmeg.

● Atole is a Mexican version of this drink, made with cornmeal and/or amaranth, with or without milk, and flavoured with various ingredients, including fruit and chocolate. One recipe is to combine 4 tbsp cornmeal, 4 tbsp ground amaranth, and a small piece of cinnamon stick with 500ml/2 cups water in a large saucepan. Bring it to the boil and simmer for 10 minutes. Add 1 litre/4 cups milk and 200g/1 cup sugar and return to the boil, stirring constantly. Pour into mugs and serve hot. Melt some chocolate into the mixture towards the end of cooking if desired. Apparently atole is considered such an important source of nourishment and comfort in the American Southwest that some hospitals make it on call for their patients.

Basic recipes

Combine these versatile recipes with the grain dishes featured in the book to make complete meals.

Grilled chicken

Use chicken breast fillets for this. Lay each one in turn between two sheets of plastic wrap on a work surface. Using a mallet or heavy rolling pin, bash along the chicken until it is evenly thin. Place in a non-corrosive container such as a Pyrex or ceramic dish and cover with 3–4 tbsp olive or vegetable oil, the juice of half a lemon (and/or some soy sauce), a crushed clove of garlic and some salt and pepper. Leave to marinate for at least an hour. To cook, heat a ridged grill-pan over a very high heat until you can see the haze rising from the surface. Lift the chicken from the marinade, lay it on the grill-pan and immediately reduce the heat to medium. Cook for 1 minute on each side – it doesn't take long because the fillets are so thin. Alternatively, cook under a hot overhead grill or salamander.

Grilled beefsteak

Lay your steaks in turn between two sheets of plastic wrap on a work surface. Using a mallet or heavy rolling pin, bash the meat until it is evenly thin. Place in a non-corrosive container. Cover with 3–4 tbsp olive or vegetable oil, the same quantity of red wine (or just a splash of red wine vinegar; for Oriental dishes use soy sauce), plus a crushed clove of garlic and some salt and pepper. Add robust herbs such as thyme, sage or rosemary if you like. Leave to marinate for at least an hour. To cook, heat a ridged grill-pan over a very high heat until you can see the haze rising from the surface. Lay the steak on the grill-pan and immediately lower the heat to medium. Cook for 1 minute, then turn and cook on the other side for another minute. The thinness of the bashed meat makes it cook quickly and therefore stay juicy. Alternatively, cook under a hot overhead grill or salamander.

Grilled prawns

You will probably want to serve around 3–5 large prawns per person. Choose raw or 'green' prawns – frozen if you like. Remove the heads, shells and tails if desired – you may only want to take off the heads and the legs under the belly. Place in a non-corrosive dish and cover with 6 tbsp olive or vegetable oil, a crushed clove of garlic, some lemon juice or soy sauce, a little chopped chilli, and some salt and pepper. Marinate for at least an hour – longer if the prawns are frozen and you are using the marinating period in tandem with defrosting. To cook, heat a ridged grill-pan over a very high heat until you can see the haze rising from the surface. Lift the prawns from the marinade, letting the excess drip away. Lay them on the grill-pan and immediately lower the heat to medium. Cook for 1½ minutes on each side. Alternatively, cook under a hot overhead grill or salamander.

Grilled fish

You can use whole fish or large fillets for this method. Clean and trim the fish as necessary. If the fish is whole, make a few diagonal slashes along each side with a sharp knife. Rub some olive or vegetable oil all over the fish and season lightly. Heat a ridged grill-pan over a very high heat until you can see the haze rising from the surface. Lay the fish on the grill-pan and immediately lower the heat to medium. Cook for 3-5 minutes on each side depending on the size and meatiness of the individual species. Alternatively, cook under a hot overhead grill or salamander.

Pan-roast fish

Choose fillets of meaty fish such as sea bass, barramundi and salmon for this method. It also suits guinea fowl. Heat the oven to 200°C/400°F/Gas mark 6. Brush an ovenproof skillet lightly with oil. Place it over a high heat until you can see the haze rising from the surface of the pan. Lightly season the fish fillets and place them in the pan skin-side down. Using a fish slice, press down firmly on the fish to help sear the skin-side, and cook for 1–2 minutes. Transfer the skillet to the hot oven and bake for 10 minutes. Alternatively, if you don't have an ovenproof skillet, you can sear the fish in a regular frying pan and transfer it to a greased baking tray before placing it in the oven.

Roast chicken

Heat the oven to 220°C/425°F/Gas 7. Line a roasting tin with a double thickness of foil. Brush the chicken all over with olive oil, then massage it with 1 tbsp or so of butter, a crushed clove of garlic and some salt and pepper. Place the chicken in the roasting tin and add 120ml/ ½ cup water. Roast for 15 minutes, then reduce the heat to 190°C/375°F/Gas 5 and roast for 25 minutes. Remove the roasting tray from the oven, leaving it on, and carefully turn the bird upside down to help it cook evenly and keep the breast meat moist. Baste it generously with the pan juices and return to the oven for a further 25 minutes. The chicken is cooked when the thighs, when pierced with a skewer, exude clear rather than bloody juices. Remove from the oven and leave to rest for 10 minutes before carving. Skim the fat from the roasting tin and use the remaining pan juices as a sauce.

Roast lamb

Heat the oven to 250°C/500°F/Gas 10 or the highest setting. Rub a small to medium leg of lamb all over with sea salt and freshly ground pepper and place it in a roasting tin. Place in the oven for 1 hour. That's it. This method from Barbara Kafka, it must be said, is controversial. If you prefer, you can roast the lamb at 230°C/450°F/Gas 8 for 20 minutes, then reduce the heat to 200°C/400°F/Gas 6 and continue cooking for a further hour or so. To calculate the timing for this method, weigh the lamb before roasting. It needs to cook for 15 minutes per 500g/1lb 2oz after that initial 20 minutes in the oven at the higher temperature. Other options include using a shoulder of lamb instead of a leg. You can also rub the meat all over with olive oil, strew it with crushed fresh sprigs of thyme or rosemary, and/or tuck some unpeeled cloves of garlic around the meat. After cooking, and setting the lamb aside to rest for 15 minutes, make a sauce from the pan juices by first skimming off the fat from the roasting tin. Place the roasting tin over a high heat and pour in about 350ml/1½ cups lamb stock, plus a slug of red or white wine if desired. Stir the pan vigorously with a wooden spoon to scrape up and incorporate the caramelized cooking juices while bringing the liquid to a vigorous boil. Boil until the volume of liquid has reduced by half, add the juices the resting lamb has exuded, and strain the sauce before serving.

Poached beef

Choose good-quality beef fillet steaks for this dish. Bring a generous quantity of beef or veal stock to the boil in a large saucepan. Add a few vegetables or flavourings if desired, but no salt. Reduce the heat so that the stock only just simmers and lay the meat in the pan. Cook at this low temperature for 3–5 minutes, or a bit longer, depending on how well done you prefer your meat and its exact size. Be aware that a more vigorous boil will cause the meat to toughen. Lift the steaks from the stock and serve as desired. The broth can be served with the steaks if required, or used in sauces, soups, stews and so on.

Deep-fried chillies

Pedrón peppers from north-western Spain are best for this (they are typically mild though statistically one in fifty is very hot). Heat a heavy pan filled around 2.5cm/1in deep with olive oil. Add the whole chillies and fry for 2–3 minutes until just starting to brown. Lift the chillies from the oil and drain on a tray covered with several thicknesses of absorbent kitchen paper. Salt the chillies generously and serve.

Roast peppers or chillies

Heat the oven to 180°C/350°F/Gas 4. Rub the peppers or chillies with olive oil and place them on a baking tray. Place in the oven for 15–20 minutes, or until just starting to brown and soften. Remove from the oven, leave to cool slightly and gently pull out the core and seeds if desired before serving. If you want to roast the peppers or chillies so that they can be peeled, place them in the oven at 200°C/400°F/Gas 6 and roast for 20 minutes or until the skins are blackened and blistered. Remove from the oven, place in a heatproof bowl and cover the bowl with plastic wrap. Leave to steam for 10–15 minutes. When cool enough to handle, peel the skins away and discard the core and seeds, but retain the cooking juices to add to your dish.

Sautéed mushrooms

Trim 120g/4oz fresh wild mushrooms as necessary and cut into large but even-sized pieces. In a large frying pan or sauté pan, heat 1 tbsp olive oil with 1 crushed garlic clove. Add the mushrooms and 1 tsp chopped parsley and cook for 3–5 minutes, stirring occasionally. When the mushrooms have released all their liquid and the liquid has evaporated, stir in ½ tbsp tomato paste, season the mixture to taste with salt and pepper and keep cooking for a further 5 minutes before serving or further use.

Indispensable tomato sauce

In a small saucepan, place 400g/1¾ cups canned, chopped plum tomatoes in natural juice. Add 2 tbsp butter or olive oil and half a small peeled onion, unsliced. Place over a low heat and leave to cook very gently, stirring occasionally, and whenever you feel like it, squashing the tomatoes against the side of the pan to make the sauce smoother. The sauce should be thick after 20–30 minutes, but if not done to your liking, keep cooking. Remove the onion. Season the sauce to taste with salt and pepper and add whatever herbs you fancy before serving or further use. For quick and easy 'baked beans', heat a drained can of small white beans such as navy or canellini in the finished sauce until piping hot.

Chicken jus

Only use very good quality chicken stock for this. In a small heavy saucepan, bring 225ml/1 cup stock to the boil and boil it very hard until it has reduced in volume by at least half. You may wish to boil it down further. The result will be a very intense but glossy clear gravy to spoon over and around chicken dishes. You can also do this in a roasting tin after roasting a chicken, setting it aside to rest and spooning off the excess fat from the tin. In this case, while bringing it to the boil on the stovetop, stir the sauce vigorously with a wooden spoon to incorporate the caramelized cooking juices from the base of the tin. Add a slug of white wine, or some sherry or brandy, while it is boiling. Adjust the seasoning to taste before serving.

Korean chilli sauce

Finely chop 1 large garlic clove and 1½ spring onions. Combine them in a small bowl with 1½ tsp rice vinegar or cider vinegar, ½ tbsp sesame oil, 1 tsp sesame seeds, 1 tsp granulated or caster sugar and mix in 1 tbsp kochujang (Korean red pepper paste). If the authentic Korean paste is unavailable, use Chinese red chilli paste.

Dashi

In a stockpot, place 2 litres/8¾ cups water and a 10cm/4in piece of kombu seaweed. Place over a moderate heat and cook for 10 minutes without letting it boil. When the kombu has fully expanded, add 30g/1oz dried bonito flakes (katsuo-bushi) and simmer gently for 5 minutes. Remove from the heat and leave the pot to stand until the fish flakes settle at the bottom. Strain the stock, discarding the solids, before further use. Reasonable instant dashi powder is available.

Coconut broth

Bring 450ml/2 cups fish stock, or vegetable or chicken stock, and 6 tbsp dry white wine to the boil in a small saucepan. Boil hard until the liquid has reduced in volume by about half. Remove from the heat and stir in 150–175ml/⅔–¾ cups good-quality coconut cream. Return to a gentle heat and warm through. Adjust the seasoning to taste with salt and pepper, plus a hint of Thai fish sauce or soy sauce if desired, and serve warm with fish and seafood dishes.

Quick berry compote

Pick over and rinse several handfuls of mixed berries and place them in a heavy pan with 1 tbsp caster sugar. Stir, then cover and cook gently over a very low heat for 10–15 minutes, stirring frequently to prevent sticking.

Toffee sauce

Combine 400g/2 cups sugar, ¼ tsp cream of tartar, a pinch of salt, and 6 tbsp water in a large, heavy saucepan. Place over a low heat and leave to cook, without stirring, for 2–5 minutes until some sugar begins to melt and turn golden. Raise the heat to medium and continue cooking, stirring occasionally, until all the sugar has melted and the caramel is a deep golden colour (or, more specifically, the 150°C/300°F 'hard crack' stage). Pour 4 tbsp cream slowly down the side of the pan, stirring constantly to combine. Transfer the sauce to a heatproof bowl and use as required.

Cream cheese frosting

In a mixing bowl, beat together 60g/¼ cup diced cream cheese, 30g/2 tbsp butter, 1 tsp vanilla extract and 1 tsp lemon or lime zest until the mixture is smooth and fluffy. Sift 1½ cups icing sugar and then gradually beat it into the cream cheese mixture until the frosting is well combined, creamy and light. This will make enough to cover the top and sides of a 23cm/9in cake.

Best-ever cheesecake filling

This makes enough to fill a 20cm/8in crust. In a mixing bowl, beat 750g/1lb 10½oz diced cream cheese and 1 tsp vanilla extract until smooth and fluffy. Set aside. In the bowl of an electric mixer, whisk 4 eggs together until thick and light, then gradually beat in 200g/1 cup caster sugar. Slowly add spoonfuls of the cream cheese mixture to the egg mixture, beating until smooth after each addition. Mix in 1 tbsp grated lemon zest and 2 tsp lemon juice. Pour the mixture into the prepared crust, and bake at 180°C/350°F/Gas 4 for 25–30 minutes. The most important bit: open the oven door and leave the cheesecake to cool right down in the oven. Only then place it in the fridge to chill.

Crumb crust

Place 225g/8oz broken crackers or biscuits in a food processor and pulse to give fine crumbs. Add 5 tbsp caster sugar and some spices if desired, then, with the machine running, pour in 125g/½ cup melted butter. Remove the dough from the processor and divide it into two or three pieces. Place a piece in the bottom of a spring-clip cake tin. Using the base of a large, heavy glass, press the dough out across the base of the tin to give a thin, firmly packed crust. Work the rest of the mixture up and around the side of the tin and fill in any gaps. Chill for 30 minutes before filling.

Rich sweet pastry

Sift 200g/1⅛ cups plain wheat flour into a mixing bowl and make a well in the centre. Dice 100g/7½ tbsp butter and add to the bowl with an egg yolk, 2 tsp caster sugar, ½ tsp salt and 3 tbsp water. Using the fingertips of one hand, work the ingredients together until they form a dough – add a little more water if necessary. Knead lightly for 1–2 minutes, then shape into a ball, wrap in plastic and chill for at least 30 minutes before rolling it out. This gives enough to line a 20cm/8in tart tin with some left over to form a pie top.

Afterword: Further fields

Here is an introduction to a range of interesting foods that, if not actually grains, can be used in ways similar to the grains and grain products featured in this book. A few are already included as possible variations of recipes. Some are hippy wholefood favourites, others have a more intriguing culinary heritage grounded in traditional regional dishes from around the world. None has hit the mass market big-time but all are worthy of some further attention as they add welcome nutritional variety to cooking and are increasingly available from fine food retailers and the better organic health stores.

Chestnut flour

As more and more speciality Italian ingredients are becoming available in other countries, chestnut flour (sometimes sold as *farina dolce* or 'sweet flour') is increasingly easy to buy. Castagnaccio is a wonderful cake from Italy's Lucchesia province made from chestnut flour, pine kernels, sultanas, rosemary and olive oil. The ancient Romans combined chestnut flour with wheat for breads and biscuits, versions of which are still made today. Historically it was also used on its own to make polenta, and can be employed as a thickener.

Chickpea flour

Flour made from chickpeas (also known as garbanzos) is a favourite ingredient of India and Italy, where it is known as besan and ceci flour respectively. In India it is made into fritter batter, snacks such as pakoda, and sweets such as laddu, as well as being used as a thickening agent for curries. The Italians use it for socca or farinata, a simple street-food pancake, also known as *Calda! Calda!*, meaning Hot! Hot! Besan and ceci flour are relatively interchangeable in cooking.

Hemp

Yes, we've all heard the jokes about *cannabis sativa*, but hemp seeds are edible and not narcotic. They are used as food in Japan, Poland and Russia, and are increasingly manufactured into wholefood products such as snack bars and ice cream.

Job's tears

Also known as adlay, job's tears are a grain, *Coix lacryma-jobi*. Perhaps I should have included a full chapter about them, but at the time of writing they are so utterly, utterly difficult to find in shops there seems no point. The unhulled grains are tear-shaped, hence the name, though once the hull is removed and polished the grains look a bit like peanuts with a wide groove running down one side. Job's tears are used in traditional Chinese medicine and thought to combat several ailments, including cancer and problems with the nervous, circulatory and digestive systems. Another common use is in the manufacture of rosaries. They are native to Southeast Asia, and most widely grown in the Philippines, however they have spread in a minor way to countries including India, Spain, Portugal and Brazil.

Linseed

Linseed, or flax, is occasionally eaten as a grain in India, however it is fast gaining popularity elsewhere as an addition to bread – not, incidentally, a new idea in Europe. In rugged, seeded baked goods it works well, and is an extremely nutritious source of fibre and essential fatty acids as well as pharmacological levels of flavonoids (powerful antioxidants), making linseeds an important addition to the diet for those who won't or can't eat soy. Some people take linseed oil as a nutritional supplement. The seeds can be ground and stirred into cereals, but both oil and meal turn rancid quickly. If you want to use it ground it is essential to grind it yourself and use it immediately. When purchasing linseed oil, always buy the smallest bottle you can find.

Lupins

Lupins are best known as the booty of Monty Python's dandy highwayman Dennis Moore. He stole from the rich and gave to the poor only to receive the complaint: 'All we've eaten mate for the last four bleeding weeks is lupin soup, roast lupin, steamed lupin, braised lupin in lupin sauce, lupin in the basket with sautéed lupins, lupin meringue pie, lupin sorbet.' It may be some time before lupins are taken seriously as a source of food, however they can indeed be used in myriad ways and make a good flour that people with wheat and gluten allergies would do well to seek out in sophisticated health food shops. An increasing number of manufacturers are including lupin flour in alternative breads and pasta products. In speciality European delicatessens you may find lupini, which are pickled or dried and salted lupin seeds that look rather like yellow broad beans. A favourite of the Romans, they are used in southern Italy, Peru and the Middle East and can be eaten as a cocktail snack or mezze dish.

Mesquite

Prosopis juliflora or *veluntina* is native to central America but under-utilized there as food, despite its excellent nutritional properties. It's better known as a source of fragrant wood for barbecues, and a plant from which bees make aromatic honey. The young green plant pods can be cooked whole as a vegetable, while the naturally sweet seeds (*péchita*) are traditionally ground into a yellowy flour or meal for flatbreads and gruel-type drinks. In *Coming Home to Eat*, Gary Paul Nabhan tells us that mesquite was a principal ingredient in the diet of American desert peoples for around ten thousand years, and that early Spanish explorers liked it, but it fell from favour during the nineteenth century. The use of mesquite for barbecues is somewhat controversial as it involves cutting down trees that would be better left standing as a source of food – rich in lysine, calcium, iron, zinc and more – for local communities.

Peasemeal

A traditional ingredient of Scotland, peasemeal is a yellow-brown flour made by roasting and then grinding yellow field peas. It is used like oat and barley meals, most simply made into brose by placing a handful in a bowl and covering it with hot milk, water or stock then stirring in a dab of butter and some salt. Flavourings such as dried fruit may be added. Peasemeal is also used in griddle-baked bannocks or scones. The traditional British dish pease pudding is made with dried peas rather than a ground meal.

Sago

As my friend Claire Clifton says, you'd tend to hope that if the cannibals of New Guinea were going to eat you, they'd have you with a nice red wine sauce, but no: the final insult is that you'd be served with sago. Used in parts of the Americas, India, Africa and throughout Southeast Asia, sago is manufactured from the stem of the sago palm, as well as some other palms, including the date palm. In countries of origin it is sold in a variety of formats, however in the West it is most commonly found shaped into beads or pearls. Sago is good used in milk puddings as well as those made from dried or fresh fruit. It can also be added to soups. An immediate hit in Britain when it was first imported in the eighteenth century, it has declined in popularity, but recently drinks of pearl sago and tea have acquired a bit of a cult following in the English-speaking world, thanks largely to their popularity in trendy Japan and Hong Kong.

Soya flour

The rich, creamy and somewhat milk-flavoured flour made from soybeans is a common sight in health food stores. Its chief purpose is adding protein, flavonoids and moistness (for it is high in fat) to baked goods requiring a gluten-free replacement for wheat flour, or as a substitute for eggs. However it also makes things leaden and if possible soy flour should not comprise more than one-third of a baked item's dry ingredients. Coeliacs aside, I don't know any real cooks of any nationality who use soya flour, and think there are too many people of puritanical bent who make using it, rather than good cooking, their goal. There are many wonderful ingredients and dishes of true culinary heritage that can be made from soybeans; this, like textured vegetable protein (TVP), isn't one of them.

Tapioca

Two things spring to mind when I think of tapioca. One is the song and dance routine based on the word from *Thoroughly Modern Millie*. The other is boys at school in England saying: 'Eeer yuk – frog's balls!' whenever we had it for lunch. Being from Australia, where school lunches were tuckshop sandwiches or sausages rolls and a vanilla slice or bar of chocolate, I thought English school dinners, where you sat down at tables for a proper two-course meal, were pretty amazing, especially when there was steamed syrup sponge pudding to be had. Despite what the lads said, the tapioca was great

too. When the English say tapioca they generally mean the milk pudding made from it, however it has other uses and the Brazilians, for example, make it into a delicious-sounding pudding with red wine or grape juice. Although it is granular, tapioca is in fact a highly processed product manufactured from a pulp that is made from the cassava root. The cassava is native to the Americas but has become an important ingredient in Asian and African cooking. Tapioca is available as small and large beads, flakes and flour, and is pretty much interchangeable with sago. Indeed, if you walk into some shops and ask for sago they will give you tapioca and tell you it's the same thing.

Urad dal

These are skinned and split black lentils, though as the skin has been removed they do not look black but creamy-grey. Freshly ground urad dal are used in the cooking of southern India, most typically for dosa batter. There was a time when any recipe requiring something to be soaked and ground in the home was immediately passed over in most Western kitchens. However I think this is slowly changing as food processors become more powerful and we increasingly appreciate the variety of exotic tastes and textures offered by foods such as lentils. In southern India urad dal are also used simply in their skinned and split form, almost like a spice, added to the oil at the beginning of making a dish along with mustard seeds, dried chillies and curry leaves. In northern India the black lentils are often used whole. The use of the word dal indicates the pulse has been split. Note that channa dal (split chickpeas), toor dal (yellow lentils), masoor dal (red lentils), and moong dal (green mung beans) are different ingredients and are not appropriate substitutes.

A brief guide to nutrition terminology

Amino acids

Often described as the 'building blocks' of proteins, amino acids are made of hydrogen, carbon, oxygen and nitrogen. Around 20 of them work together to produce protein in the body. Most amino acids are made by the body, however eight of them have to be taken in through diet. These are known as essential amino acids and are: isoleucine, leucine, lysine, methionine, phenylalanine, threonine, tryptophan and valine. Children and infants require an additional essential amino acid, histidine. It is important that the essential amino acids are supplied in the correct ratios to produce the body's protein. Although this seems complicated, it is simply a matter of eating various food sources of essential amino acids. For more on this see Protein.

Antioxidants

Antioxidants are a group of vitamins, minerals and phytonutrients that prevent or retard oxidization. On a practical level this means they stop (or delay) foods, especially fatty foods and oil, becoming rancid. In the body, antioxidants retard the breakdown of tissues. It is thought that they have a particular role in preventing heart disease, cancer, dementia and skin damage or deterioration. Wholegrains are important sources of antioxidants, while pulses and vegetables are even more so.

Carbohydrates

The World Health Organization recommends that 55 to 75 per cent of calories come from carbohydrates, which are classified in various ways. All foods except oils contain some degree of carbohydrate, however some are substantially richer in carbohydrates than others. Complex carbohydrates are highly beneficial to health and can be found in starchy foods including grains, potatoes, pulses, vegetables and a few starchy fruits such as bananas. Complex carbohydrates may be unrefined, such as wholegrains, or refined, as in refined white rice and white wheat flour. Unrefined or marginally refined carbohydrates are preferable as they have a high content of fibre and other nutrients (see Fibre). Intrinsic sugars are another type of carbohydrate that is found naturally in the cells of most fruits, as well as in vegetables. Extrinsic sugars are refined by manufacture or processing and include such ingredients as table sugar, honey and syrups. Although it is considered important to consume a high level of carbohydrates, intrinsic sugars are preferable because extrinsic sugars provide substantially less nourishment per calorie than intrinsic sugars, and are associated with tooth decay.

Cholesterol

Cholesterol can be produced by the body or taken in through food. The first type is known as serum or blood cholesterol and is produced by the liver. Far from being a bad thing, it is essential to life. Its purposes include helping to build hormones, produce vitamin D, and protect nerve fibres. The cholesterol taken into the body through food is termed dietary cholesterol and is not essential. Foods high in dietary cholesterol also tend to be high in fat.

Fat and fatty acids

Fats are essential nutrients. Each type of fat is comprised of a number of chains known as fatty acids. Fatty acids are made of carbon, hydrogen and oxygen linked together. There are several types, which may be classified as saturated, mono-unsaturated, or polyunsaturated. Current dietary advice is to limit overall fat intake and, when fat is eaten, to favour foods that are high in mono- and polyunsaturated fats while keeping saturated fat intake low.

Most fats can be produced by the body however there are two important groups of polyunsaturated fats, Omega-3s and Omega-6s, that are helpful to consume through food. These are known as essential fatty acids and it is thought that they may help prevent or control a wide variety of nutrition-related ailments. Omega-3 fatty acids – which are found in oily fish, dark green leafy vegetables, nuts and seeds, and some oils – are thought to be beneficial in reducing the risk of heart disease, preventing the growth of tumours, lowering blood pressure and building immunity. Omega-6 fatty acids are thought to be especially helpful to women with premenstrual syndrome or menopausal symptoms and are found in a variety of seeds and nuts, and oils made from them. Trans fats are unsaturated fats that have been altered, usually via food processing, so that they are hard at room temperature. They subsequently act like saturated fats in the body, and may even be more damaging. It is therefore important to keep consumption of trans fats low. They are typically found in margarine and mass-produced baked goods, even those such as crackers that are generally perceived as healthy foods.

Fibre

Fibre and 'roughage' are common terms for non-starch polysaccharides, components of food that pass through the body undigested. Although our bodies do not absorb fibre, it is an essential group of nutrients. Traditionally known for the role it plays in the digestive system (it 'keeps you regular', preventing constipation and haemorrhoids), it is also vital in weight and hunger control, balancing blood sugar levels, reducing cholesterol and preventing diabetes, bowel cancer and irritable bowel syndrome. Grains, especially wholegrains, plus pulses, vegetables and fruit are good sources and eating a variety of these foods helps to ensure that you consume the different types of fibre necessary for good health. Soluble fibre is gel-like and dissolves in water. It is especially associated with lowering levels of disadvantageous LDL cholesterol in the blood. There are various types of soluble fibre: beta-glucans is found in oats, barley and rye; arabinose is found in pulses; and pectin is found in citrus fruits and apples. Insoluble fibre does not dissolve in water. It is bulky and particularly associated with moving food through the body and making you feel full after eating. Wheat, corn and rice, along with pulses and vegetables, are important sources. Like insoluble fibre, resistant starch also passes through the body undigested and helps provide bulk. It is found in cooked grains and potatoes.

Flavonoids

Flavonoids are a group of compounds that work as antioxidants and phytoestrogens. They are particularly associated with the prevention of various cancers and include flavonols, flavones, isoflavones and bioflavonoids. Linseeds, soy products, pulses, onions, apples, cranberry juice and tea are good sources.

Gluten

Gluten is a protein found in some grains and flours made from them. It plays a key role in baking as it helps give a light but cohesive texture. In bread making gluten forms long, elastic strands that trap the gases produced by the yeast, creating a good rise. Coeliac disease is an inflammatory disease of the gastrointestinal tract caused by an allergy to gluten, which damages the sufferer's intestinal lining and prevents absorption of nutrients.

Lignans

A type of phytochemical (see Phytochemicals), lignans are a fibrous component of plant cell walls that may help prevent cancer. Wholegrains are an important source of lignans, as are linseeds, berries and some vegetables.

Linoleic acid

Linoleic acid is one of the Omega-6 group of essential fatty acids (see Fat and fatty acids).

Lysine

Lysine is one of the essential amino acids that must be consumed through diet (see Amino acids). Some cereal grains are low in lysine; other grains covered in this book are by contrast notably high.

Nutrients

Nutrients are the components of food that nourish the body and maintain life. The categories traditionally known as essential nutrients are carbohydrates, protein, fat, vitamins, minerals and water. However scientists are constantly discovering other key food components, such as phytochemicals, that are also important to good health.

Phytochemicals

Phytochemicals are components of plant foods that actively protect health. The use of the prefix phyto indicates that they are derived from plants however not all phytochemicals contain this prefix (for example bioflavonoids and rutin). There are thousands of types of phytochemicals. Phytoestrogens (of which isoflavones and lignans are two main types) are structured similarly to the female hormone oestrogen and seem to be helpful in preventing hormone-dependent cancers of the breast and womb, as well as heart disease and osteoporosis. Wholegrains are important sources of lignans and therefore phytoestrogens. Phytosterols are also found in grains (particularly the oils) and work to reduce cholesterol in the body, while phytic acid, located in the bran, seems to protect against some cancers.

Protein

Protein is one of the macronutrients required by the body for growth, maintenance and repair. The word comes from a Greek word meaning 'of prime importance'. Although very important, it is not necessary to eat a great deal of it, though the optimum amount tends to be hotly debated by nutrition experts. In order to consume the essential amino acids needed for the body to produce protein, you need to eat foods that are sources of protein. There are two groups. 'Complete proteins' are able to provide all eight essential amino acids on their own and include soybeans, meat, fish and poultry, and dairy products such as eggs and cheese. Foods classed as 'incomplete proteins', including grains, nuts, pulses and legumes, have some essential amino acids, but not others, and the exact ones they contain vary with each individual food. Nevertheless they are arguably healthier than some complete proteins and it is important to general health to consume them. They form complete proteins in the body when eaten in combination with other incomplete proteins, for example as lentils and rice, or beans on toast. Recent studies show that combining incomplete protein foods does not necessarily need to be done in the same meal, simply throughout the course of each day's eating.

Cooking guide

Grain – per 225ml/1 cup	Stovetop liquid	Stovetop boiling time	Pressure-cooking liquid	Pressure-cooking time
Amaranth	3 cups	20 mins	2 cups	15–20 mins
Barley, semi-pearled or pot	3 cups	45–75 mins	2¼ cups	45 mins
Barley, white-pearled	2½–3 cups	20–30 mins	2¼ cups	15–20 mins
Barley, grits or cracked	3 cups	20 mins	2¼ cups	15 mins
Buckwheat, groats	2 cups	12–20 mins	n/a	n/a
Corn, coarse meal or polenta	3–4 cups	35–40 mins	2½ cups	20 mins
Corn, grits	3–4 cups	30 mins	2½ cups	20 mins
Corn, hominy or samp	3–4 cups	1½ hours	2½ cups	40–50 mins
Corn, posole large	4 cups	2–3 hours	3 cups	50 mins
Millet, hulled	3 cups	15–20 mins	2½ cups	8–10 mins
Oats, wholegrain	4 cups	60–75 mins	2½ cups	50 mins
Oatmeal, coarse or steel-cut	4 cups	45 mins	3½–4 cups	30 mins
Oatmeal, medium	5 cups	30 mins	n/a	n/a

Sensory cues	Yield
Grains meld into a thick and gluey porridge	3 cups
Grains expand to give a texture like crunchy, juicy berries	4¼ cups
Grains swell to give a texture like juicy pillows	3½ cups
Grains are swollen, juicy and crunchy with an uneven, gritty texture	3 cups
Grains swell and soften, but retain texture of angular pebbles	2½ cups
Tiny starch grains burst to give a blowsy, textured, thick mixture	3–4 cups
Grits become a thick, creamy paste	3–4 cups
Grains soften gradually to textured paste	2½ cups
Grains burst open like a flower and are softly chewy	2½ cups
Grains puff up and soften to give a fine, light sandy texture	3½ cups
Grains maintain their shape, but become tender and creamy	3 cups
Grains gradually meld into a thick paste with a hint of crunch	3 cups
Grains gradually meld into a thick, slightly textured paste	3 cups

Nutritional facts

Grain – per 100g (3½oz), uncooked	Calories	Carbohydrate (g)
Amaranth	391	63.1
Barley, wholegrain	301	64
Barley, semi-pearled or pot	346	76.4
Barley, white-pearled	360	83.6
Barley, meal	345	74.5
Buckwheat, groats	343	71.5
Buckwheat flour, wholemeal	335	70.6
Cornmeal, wholemeal yellow	362	76.9
Corn grits	371	79.6
Cornmeal, masa, enriched yellow	365	76.3
Cornmeal, fine white degermed	366	77.7
Millet	378	72.8
Millet flour	354	75.4
Oats, wholegrain	389	66.3
Oatmeal, medium	401	72.8
Quinoa	374	68.9

Sources: Various
N = component is present in significant quantities, but figures are unreliable.

Fibre (g)	Protein (g)	Fat (g)	Gluten
2.9	15.3	7.1	No
17.3	10.6	2.1	Yes
10.4	10.7	2.2	Yes
5.9	7.9	1.7	Yes
10.1	10.5	1.6	Yes
10	13.2	3.4	No
10	12.6	3.1	No
7.3	8.1	3.6	No
1.6	8.8	1.2	No
9.6	9.3	3.8	No
7.4	8.5	1.6	No
8.5	11	4.22	No
N	5.8	1.7	No
10.6	16.9	6.9	Yes
6.3	12.4	8.7	Yes
5.9	13.1	5.8	No

Nutritional facts

Grain – per 100g (3½oz), uncooked	Calories	Carbohydrate (g)
Rice, polished	361	86.8
Rice, wholegrain	357	81.3
Rye, wholegrain	335	69.8
Rye flour, dark	324	68.7
Rye flour, light	367	80.2
Triticale, wholegrain	336	72.1
Triticale flour, wholemeal	338	73.1
Wheat, bulgur	342	75.9
Wheat, freekeh wholegrain	345	72
Wheat, common wholegrain	335	72.7
Wheat, petit épautre wholegrain	375	74.9
Wheat, Kamut wholegrain	359	68.2
Wheat, pearled such as Pasta Wheat	337	70.4
Wheat flour, refined	341	77.7
Wheat flour, wholemeal	310	63.9
Wild rice	365	74.3

Sources: Various
N = component is present in significant quantities, but figures are unreliable.

Fibre (g)	Protein (g)	Fat (g)	Gluten
2.2	6.5	1	No
3.8	6.7	2.8	No
14.6	14.8	2.5	Yes
22.6	14	2.6	Yes
14.6	8.4	1.4	Yes
N	13	2.1	Yes
14.6	13.2	1.8	Yes
18.3	12.3	1.3	Yes
16.5	12.6	2.7	Yes
2.1	12.3	1.9	Yes
7.7	11.8	2.9	Yes
1.8	17.3	2.6	Yes
6.7	11.3	1.1	Yes
3.6	9.4	1.3	Yes
8.6	12.7	2.2	Yes
7	13.2	1.7	No

Sprouting

Grain	Soak	Rinse	Days	Length of sprout
Barley	6–10 hours	3 times per day	2–3 days	0.5cm/¼in
Buckwheat	15 minutes	every hour for 4 hours then 3 times per day	2–3 days	1–2.5cm/½–1in
Millet	5–7 hours	3 times per day	2 days	3mm/⅛in
Oat	3–5 hours	2 times per day	2 days	0.5cm/¼in
Quinoa	2–4 hours	3 times per day	2–4 days	0.5–3cm/¼–1¼in
Rice	6–10 hours	3 times per day	2–3 days	3mm/⅛in
Rye	6–10 hours	2 times per day	2–3 days	0.5–1cm/¼–½in
Triticale	6–10 hours	2 times per day	2–3 days	0.5–1cm/¼–½in
Wheat	6–10 hours	2 times per day	2–3 days	0.5–1cm/¼–½in

See page 111 for sprouting instructions.
Chart is for indoor jar or tray sprouting, not for the production of wheatgrass.
In all cases it is essential to purchase wholegrains specifically for sprouting from a specialist supplier.
Adequate drainage is essential to prevent mould growth.
Grains are unlikely to sprout successfully in hot weather.

Adapted from chart by *Sprouting Publications*.

Index

Bibliography

American Heritage Magazine, *The American Heritage Cookbook*, Simon and Schuster

Anderson, E.N, *The Food of China*, Yale University Press (New Haven & London, 1988)

Booth, Shirley, *Food of Japan*, Grub Street (London, 2000)

Braunstein, Mark M. *Sprout Garden*, Book Publishing Co. (Summertown, 1999)

Brown, Catherine, *Scots Cookery*, Richard Drew Publishing (Glasgow, 1989)

Carluccio, Antonio, *An Invitation to Italian Cooking*, Pavilion Books (London, 1991)

Christian, Glynn, *Edible France*, Grub Street (London, 1996)

Christian, Glynn, *New Delicatessen Food Handbook*, Good Food Retailing Publications (Caterham)

Cooper, Katharine, *The Australian Triticale Cookery Book*, Savvas Publishing, (Adelaide, 1985)

Del Conte, Anna, *The Classic Food of Northern Italy*, Pavilion Books (London, 1995)

Devi, Yamuna, *The Art Of Indian Vegetarian Cooking*, Century Hutchinson (London, 1990)

Dupleix, Jill, *New Food*, William Heinemann (Melbourne, 1994)

Eyre, David & the Eagle cooks, *Big Flavours & Rough Edges*, Headline Book Publishing (London, 2001)

Farmer, Fannie, *The 1896 Boston Cooking School Cookbook*, Gramercy Books (New York)

Field, Carol, *In Nonna's Kitchen*, HarperCollins (New York, 1997)

Flower, Barbara and Rosenbaum, Elisabeth, *The Roman Cookery Book*, a Critical Translation of The Art of Cooking by Apicius, Harrap (London, 1958)

Gonzalez, Angeles (ed), *Maize Cookery* (Mexico, 1993)

Gould, Kevin, *Dishy*, Hodder and Stoughton (London, 2000)

Granger, Bill, *Sydney Food*, Murdoch Books (Sydney, 2000)

Grigson, Jane, Food With The Famous, Grub Street (London, 1991)

Hambro, Nathalie, *Particular Delights*, Jill Norman & Hobhouse (London, 1981)

Harris, Jessica B, *The Africa Cookbook*, Simon and Schuster (New York, 1998)

Hartley, Dorothy, *Food In England*, Little, Brown and Company (London, 1996)

Hauser, Susan Carol, *Wild Rice Cooking*, The Lyons Press (New York, 2000)

Helou, Anissa, *Lebanese Cuisine*, Grub Street (London, 1994)

Hensperger, Beth, *Baking Bread: Old and New Traditions*, Chronicle (San Francisco, 1992)

Jerome, Carl, *Cooking For A New Earth*, Henry Holt and Company (New York, 1993)

Kiple, Kenneth, and Ornelas, Kriemhild Conee (ed), *The Cambridge World History of Food*, Volume I, Press Syndicate of the University of Cambridge (Cambridge, 2000)

Koffman, Pierre, *Memories of Gascony*, Pyramid Books (London, 1990)

Lawrence, Sue, *Scots Cooking*, Headline (London, 2000)

Luard, Elisabeth, *European Peasant Cookery*, Bantam Press (London, 1986)

McLaughlin, Michael, *Good Mornings*, Chronicle (San Francisco, 1996)

Meyerowitz, Steve, *Wheatgrass: Nature's Finest Medicine*, Sproutman (Great Barrington, 1999)

Millon, Marc and Kim, *Flavours of Korea*, Andre Deutsch (London, 1991)

NSW Public School Cookery Teachers Association, *The Commonsense Cookery Book*, Angus and Robertson (Sydney 1970)

Owen, Sri, *The Rice Book*, Frances Lincoln (London, 1998)

Ortiz, Elisabeth Lambert, *Mexican Cooking*, Grub Street (London, 1998)

Oxford Symposium on Food and Cookery Proceedings, *Staple Foods,* Prospect Books (London, 1990)

Palmer, Adam, *Champneys Cookbook*, Ward Lock (London, 1999)

The Picayune, *The Picayune Creole Cook Book*, Dover Publications (New York, 1971)

la Place, Viana, *Verdura: Vegetables Italian Style*, William Morrow (New York, 1991)

Reader's Digest, *Complete Guide To Cookery*, Reader's Digest (London,1995)

Reader's Digest, *Low Fat No Fat Cook Book*, Reader's Digest (London,1998)

Reader's Digest, *Your Cookery Questions Answered*, Reader's Digest (London,1996)

Roden, Claudia, *The Book Of Jewish Food*, Penguin Group (London,1997)

Rodriguez, Douglas, *Latin Ladles*, Ten Speed Press (Berkeley,1997)

Rodriguez, Douglas, *Nuevo Latino*, Ten Speed Press (Berkeley,1995)

Rose, Peter (ed), *The Sensible Cook: Dutch Foodways in the Old and the New World* (1989)

Rosengarten, David, with Dean, Joel and DeLuca, Giorgio, *The Dean and DeLuca Cookbook*, Random House, (New York, 1996)

Salaman, Rena, *Healthy Mediterranean Cooking*, Stewart, Tabori & Chang (New York,1996)

Shaida, Margaret, *The Legendary Cuisine Of Persia*, Grub Street (London, 2000)

Shulman, Martha Rose, *The Bread Book*, Macmillan (London,1990)

Sinclair, Ellen (ed), *Australian Women's Weekly Original Cookbook*, Golden Press (Silverwater, 1970)

Slater, Nigel, *Appetite*, Fourth Estate (London,2001)

Snyman, Lannice. Sawa, Andrzej, *Rainbow Cuisine*, S&S Publishers (Hout Bay, 2000)

Sokolov, Raymond, *With the Grain*, Alfred A Knopf (New York, 1996)

Spencer, Colin, *Vegetarianism: A History*, Grub Street (London, 2000)

Spieler, Marlena, *Mediterranean Cooking The Healthful Way*, Prima Publishing (Rocklin,1997)

Sreedharan, Das, *Fresh Flavours of India*, Conran Octopus (London,1999)

Sreedharan, Das, *The New Tastes of India*, Headline (London, 2001)

Stobart, Tom, *The Cook's Encyclopaedia*, Grub Street (London, 1998)

Stucchi, Emanuela, *Italian Vegetarian Cooking*, Pavilion (London, 1994)

Taruschio, Ann & Franco, *Leaves from The Walnut Tree*, Pavilion Books (London, 1993)

Taylor Simeti, Mary. *Sicilian Food*, Grub Street (London, 1999)

United Nations Food and Agriculture Organisation, *Sorghum and Millets in Human Nutrition*, United Nations (1995)

Uvezian, Sonia, *Recipes and Rememberances from an Eastern Mediterranean Kitchen*, University of Texas, (Texas, 1999)

Veldsman, Peter, *Flavours of South Africa*, Tafelberg (Cape Town, 1998)

Vennum, Thomas Jr. *Wild Rice and the Ojibway People*, Minnesota Historical Society (St.Paul, 1988)

Vogue Entertaining Cookbook, *Morning Noon and Night*, Vogue Australia (Sydney, 1999)

Wells, Patricia, *The Paris Cookbook*, Kyle Cathie (London, 2001)

Williams, Patrick, *The Caribbean Cook*, Penguin Books (London, 2001)

Wing, Lorna, *Party Food*, Conran Octopus (London,1998)

Wolfert, Paula, *Couscous and other good food from Morocco*, Harper & Row (New York, 1973)

Wood, Rebecca, *Quinoa The Supergrain*, Japan Publications (Tokyo,1989)

Wood, Rebecca, *The Splendid Grain*, William Morrow (New York, 1997)

Acknowledgments

Note for Australian readers

The cup measures used here are an American 8fl oz cup, which equals approximately 225ml. Also, tablespoons are 15ml in volume.

Thanks to: Claire Clifton, Clarissa Hyman, John Campbell, Marion Moisy, Caroline Proud, Lorraine Dickey, Leslie Harrington, Lucy Gowans, Muna Reyal, Victoria Burley, Katey Day, Catharine Snow, Lyn Hall, Vanessa Courtier, Angela Bodgiano, Jason Lowe, Sue and Pat Lawrence, Emi Kazuko, Martin Bumpsteed, Keith Davidson, Brenda Houghton, Ian Fenn, Ian Shaw, Guy Dimond, Susie Theodorou, Sandra Diamond, Christine Jeffries, Linda Barcan, Bernadette Lara, Fiona Donnelly, Annette Kesler, Peter Veldsman, Johan Odendaal, Rosemary Barron, Miriam Polunin, Geoff Tansey, Marc Millon, Marlena Spieler, Colin Spencer, Daphne Lambert, Deh-ta Hsiung, Robert Pryor, Charlie Trotter, Mary Hegeman, Mark Signorio, Paul Gayler, Tim Lang, David Gibson, Margaret Shaida, Anissa Helou, David Eyre, Shaun Hill, Lorna Wing, Kevin Gould, Linda Collister, Das Sreedharan, Luke Mangan, Joseph Sponzo, Maria José Sevilla, Sarah Chase, Carlos Vargas, Paul Gayler, Helene Cuff, Caroline Stacey, Catherine Turner, Sue Style, Nancy Jenkins, Sharyn Storrier Lyneham and staff at Vogue Entertaining and Travel, the Guild of Food Writers' Sparklist, Stuart Ninian, Tony Lutfi, Kath Cooper, Rhonda Sweetgrass, Rebecca Geraghty, Philip Chamberlain, Jeremy Blower, Alison Brown, Nobuko Yamamoto, Yoshio Miyazawa, plus David Virgo and all the McCleans.

The publisher would like to thank the following photographers, agencies and architects for their kind permission to reproduce the photographs in this book:

7 Anthony Blake Photo Library (Clare Parker);23 Robert Harding Picture Library (Duncan Maxwell); 24 Cephas (Walter Geiersperger); 36 Cephas; 40 Jan Baldwin; 56 Cephas (Mick Rock); 62 Anthony Blake Photo Library (Kieran Scott); 69 Cephas.

First published in 2002 by
Conran Octopus Limited
a part of Octopus Publishing Group
2–4 Heron Quays, London E14 4JP
www.conran-octopus.co.uk

Text copyright © Jenni Muir 2002
Recipes copyright © Jenni Muir 2002
Book design and layout copyright
© Conran Octopus 2002

Publishing director: Lorraine Dickey
Senior editor: Muna Reyal
Copyeditor: Marion Moisy

Creative manager: Lucy Gowans
Designer: Vanessa Courtier, Victoria Burley
Photography: Jason Lowe
Home economist: Angela Bodgiano, Lyn Hall
Stylist: Róisín Nield

Senior picture researcher: Rachel Davies

Senior production controller: Manjit Sihra

British Library Cataloguing-in-Publication Data.
A catalogue record for this book is available from the British Library.

ISBN 1 84091 073 9

Printed in China